Managing
Oak Forests
in the
Eastern
United States

Managing
Oak Forests
in the
Eastern
United States

Edited by
Patrick D. Keyser
University of Tennessee
Center for Native Grasslands Management
Knoxville, USA

Todd Fearer
Appalachian Mountains Joint Venture
Blacksburg, Virginia, USA

Craig A. Harper
University of Tennessee
Department of Forestry, Wildlife, and Fisheries
Knoxville, USA

CRC Press
Taylor & Francis Group
Boca Raton London New York

CRC Press is an imprint of the
Taylor & Francis Group, an **informa** business

CRC Press
Taylor & Francis Group
6000 Broken Sound Parkway NW, Suite 300
Boca Raton, FL 33487-2742

First issued in paperback 2021

Version Date: 20151103

ISBN-13: 978-1-03-209809-8 (pbk)
ISBN-13: 978-1-4987-4287-0 (hbk)

Visit the Taylor & Francis Web site at
http://www.taylorandfrancis.com

and the CRC Press Web site at
http://www.crcpress.com

Contents

Preface... vii
Editors.. xi
Contributors... xiii

Section I: Background and biology: Setting the stage

Chapter 1   The mighty oak: Why do we manage oaks?........................ 3
             Bryan Burhans, Patrick D. Keyser, Craig A. Harper, and
             Todd Fearer

Chapter 2   History of eastern oak forests 7
             Marc D. Abrams

Chapter 3   Description and distribution of oak forests across the
             Eastern United States.. 19
             Christopher M. Oswalt and Matthew Olson

Chapter 4   Oak forests' value to wildlife...................................... 41
             Todd Fearer

Chapter 5   Silvics of oaks: Their biology and ecology..................... 51
             Thomas R. Fox and Jerre Creighton

Chapter 6   Oak regeneration challenges....................................... 63
             Jeffrey W. Stringer

Section II: Silviculture: What is in the tool box?

Chapter 7   Regenerating oak stands the "natural" way.................... 75
             Callie Jo Schweitzer, Greg Janzen, and Dan Dey

Chapter 8   Artificial regeneration..85
             David S. Buckley and Victor L. Ford

Chapter 9   Fire in the oak woods: Good or bad? 99
             Craig A. Harper and Patrick D. Keyser

Chapter 10 Competition control for managing oak forests 131
             Victor L. Ford, Jim H. Miller, and Andrew W. Ezell

Chapter 11 Intermediate treatments 153
             Wayne K. Clatterbuck

Section III: Managing oaks: How do I make it work for me?

Chapter 12 Where am I? 167
             Tamara Walkingstick, Kyle Cunningham, and Jon Barry

Chapter 13 What are my management objectives? 183
             Daniel Yaussy, Patrick D. Keyser, and Bryan Burhans

Chapter 14 How do I manage upland oaks?.............................193
             *David Wm. Smith, James E. Johnson, John W. Groninger,
             and Mark E. Banker*

Chapter 15 How do I manage for woodlands and savannahs? 223
             *Patrick D. Keyser, Craig A. Harper, McRee Anderson,
             and Andrew Vander Yacht*

Chapter 16 Managing bottomland oaks 247
             Mike Staten and Wayne K. Clatterbuck

Chapter 17 Managing deer impacts on oak forests.............261
             David deCalesta, Roger Latham, and Kip Adams

Chapter 18 How do I get started?...279
             Larry A. Tankersley

Index 283

Preface

Although there may be books in existence that address various aspects of the silvics or ecology of eastern hardwood forests and those that address some aspects of oak ecology, this book was prepared to address a gap in the literature regarding oaks. Specifically, the vision for this book came from a recognition by one of our coauthors, Bryan Burhans, that little practical information was available to landowners, wildlife biologists, soil conservationists, or land managers, in any single place. His recognition of the importance and the need for active stewardship of our oak forests inspired him to try to create something like *Managing Oak Forests in the Eastern United States*, which became a reality.

Although the development of this book took some time to unfold, we eventually brought together a team of 33 scientists, educators, and practitioners who collectively represent more than 8 centuries of experience in natural resources, notably hardwood forest management, and especially that pertaining to oak forests in the Eastern United States. Those involved have, in many cases, developed national and international reputations as experts in their respective areas. Others have, through decades of practical experience, developed a savvy and familiarity with the art of oak forest management that is an indispensable complement to the science. Among those actively engaged in the stewardship of oak forests in recent decades, we are hard pressed to think of how we could have improved upon this team of professionals.

The oak forests of the Eastern United States represent one of the great natural treasures of the Earth. Perhaps they are not as dramatic as the Rocky Mountains, or as mysterious as the Amazon Basin, or as exotic as the East African Plains, but they are remarkable in their own right. As one of the largest temperate, deciduous forests on the planet, they are home to thousands of species of insects, birds, amphibians, reptiles, mammals, plants, fungi—and millions of people. They are the source of millions of board feet of some of the most sought after hardwood lumber on the planet. Millions of tons of carbon dioxide are fixed, sequestered, and exchanged for oxygen by this "living lung." Billions of gallons of clean drinking water are protected by the deep root systems that clothe the millions of acres of hillsides from Appalachia to the Ozark Highlands. And millions of hours of recreation in the form of deer, grouse, turkey, and squirrel hunting, not to mention mushroom hunting, photography, and bird watching, are provided by these extensive forests. They are truly a magnificent gift to us from the Creator.

To the casual observer though, who may admire their autumn beauty, abundant wildlife, high-quality lumber, or overall productivity, the state of their underlying ecological health may be much less apparent. Indeed, that same casual observer may miss the almost complete lack of oak regeneration; or the "mesophication" of these forests; or not recognize that important, fire-adapted communities, which have been a part of this vast ecosystem for millennia, are now all but gone; or that a number of wildlife species that once called these forests home are now rarely seen or heard among their canopies because of slow, gradual changes that have gone unnoticed by so many, even those who should know better; or that vast stretches of these forests are now aging to the point where they can no longer replace themselves, to the point where their future as oak forests is in doubt; or the severe damage to understories and regeneration caused in many areas by deer overpopulation that has persisted for far too many years; and they may not see some not-so-visible forest pests that could cause cataclysmic damage to these forests in years to come (think chestnut blight and Dutch elm disease, pathogens that have robbed our oak ecosystems of these once common species).

Each of these issues is, to some degree, the result of decades of altered disturbance regimes, an ecologist's term for those natural agents that maintain healthy ecosystems. In the case of oak forests, they thrive to the extent that canopies are periodically opened to allow light to reach the forest floor, or that understories are reinvigorated by periodic, low-intensity fires. Although man has inexorably altered the pattern of such disturbances, good stewardship mandates that we use the tools of modern forestry, derived from decades of science and practice, to ensure that this great treasure is properly cared for. The timeless wisdom of Genesis (2:15) comes to mind: "The Lord God took the man and put him in the Garden of Eden *to work it and take care of it*." A reasonable person may question if an oak forest in the Eastern United States is the "Garden of Eden," but they should not question the obvious moral imperative to "take care of it"!

We have prepared this book to help those who have that stewardship responsibility more than most—landowners and managers. It is organized into three sections. The first, "Background and Biology: Setting the Stage," contains six chapters intended to provide readers with a solid, basic understanding of how this thing called an *Eastern Oak Forest* works. The second, "What Is in the Tool Box?" contains five chapters intended to give readers some practical understanding of how management can be implemented in eastern oak forests—to become familiar with the tools available and how they can use those tools. Finally, the third section, "Managing Oak Forests: How Do I Make It Work for Me?" will enable readers to clarify their objectives and chart a course to bring about the desired outcomes for the forests they are managing. We also hope that this book will help readers to recognize that the stewardship of eastern oak forests will not be successful if they take the approach that these forests can simply be ignored, protected from disturbance, or by allowing "nature to take its course." As mentioned, altered natural forces have led to altered forests and altered successional trajectories. Now and in the future, good stewardship will require wise, well-informed, and active management.

While our primary concern in assembling this book was to empower managers, we recognize that we have also created a book that can have great value as a text for college courses in forestry and natural resources or serve as a valuable reference book for anyone interested in forests and their ecology and stewardship. Our real hope though, is that those same oak forests that have benefited us in so many ways, and which we have enjoyed hunting in, walking through, studying, and being stewards of, will be there for the next generation—and those who will come after them. Perhaps this book will motivate and equip landowners and managers to take the step of educating themselves, of moving good intentions for management into real and effective action.

Editors

Patrick D. Keyser, PhD, is professor and director of the Center for Native Grasslands Management, University of Tennessee (UT), Knoxville, where he has been since 2006. Prior to UT, Dr. Keyser was a senior wildlife biologist for MeadWestvaco, a global forest products company, based out of West Virginia. He also worked as a biologist for the Virginia Department of Game and Inland Fisheries. In all these roles, he has been actively engaged in forest management, forest and wildlife research, and working with various forest landowners (private, state, federal, industry) to improve forest productivity and sustainability. Much of his current work focuses on the restoration and management of oak woodlands and savannahs and shortleaf pine communities.

Todd Fearer, PhD, is the coordinator for the Appalachian Mountains Joint Venture (AMJV) in Blacksburg, Virginia, a regional partnership among state and federal agencies, conservation organizations, universities, and industries that work to conserve habitats for migratory birds in the Appalachian Mountains. Dr. Fearer joined the AMJV in 2010 as the science coordinator, and became coordinator in 2012. Prior to working with the AMJV, he was an assistant professor at the University of Arkansas School of Forest Resources. He received a BS in wildlife and fisheries science with minors in forest science and international agriculture from Pennsylvania State University, State College. He then went on to earn both an MS and PhD in wildlife science from Virginia Tech, conducting research as part of the Appalachian Cooperative Grouse Research Project and evaluating population–habitat relationships of forest breeding birds at multiple spatial and temporal scales, respectively.

Craig A. Harper, PhD, is a professor of wildlife management and the extension wildlife specialist at the University of Tennessee where he develops wildlife management programs for the UT Extension and assists natural resource professionals with issues concerning wildlife management.

Dr. Harper maintains an active research program that specializes in upland habitat management, including the effects of silviculture, prescribed fire, and herbicide applications on habitats for various wildlife species. Dr. Harper is a Certified Wildlife Biologist and a Certified Prescribed Fire Manager.

Contributors

Marc D. Abrams
Department of Ecosystem Science
and Management
Pennsylvania State University
University Park, Pennsylvania

Kip Adams
Quality Deer Management
Association
Knoxville, Pennsylvania

McRee Anderson
The Nature Conservancy
Fayetteville, Arkansas

Mark E. Banker
Appalachian Forest Consultants–
Wildlife Services
Lemont, Pennsylvania

Jon Barry
Arkansas Forest Resources Center
Southwest Research and Extension
Center
University of Arkansas
Hope, Arkansas

David S. Buckley
Department of Forestry, Wildlife,
and Fisheries
University of Tennessee
Knoxville, Tennessee

Bryan Burhans
Pennsylvania Game
Commission
Harrisburg, Pennsylvania

Wayne K. Clatterbuck
Department of Forestry, Wildlife,
and Fisheries
University of Tennessee
Knoxville, Tennessee

Jerre Creighton
Virginia Department of Forestry
Charlottesville, Virginia

Kyle Cunningham
Arkansas Forest Resources
Center
University of Arkansas
Little Rock, Arkansas

David deCalesta
Halcyon-Phoenix Consulting
Crossville, Tennessee

Dan Dey
Forest Service–Northern
Research Station
U.S. Department of Agriculture
University of Missouri
Columbia, Missouri

Andrew W. Ezell
Department of Forestry
Mississippi State University
Starkville, Mississippi

Victor L. Ford
Southwest Research and Extension
 Center
University of Arkansas
Hope, Arkansas

Thomas R. Fox
Department of Forest Resources
 and Environmental
 Conservation
Virginia Polytechnic Institute and
 State University
Blacksburg, Virginia

John W. Groninger
Department of Forestry
Southern Illinois University
Carbondale, Illinois

Greg Janzen
Stevenson Land Company
Scottsboro, Alabama

James E. Johnson
College of Forestry
Oregon State University
Corvallis, Oregon

Roger Latham
Continental Conservation
Rose Valley, Pennsylvania

Jim H. Miller
Forest Service–Southern
 Research Station
U.S. Department of
 Agriculture
Auburn University
Auburn, Alabama

Matthew Olson
Missouri Department of
 Conservation
Forest Systems Field Station
West Plains, Missouri

Christopher M. Oswalt
Forest Inventory and Analysis
Forest Service–Southern
 Research Station
U.S. Department of Agriculture
Knoxville, Tennessee

Callie Jo Schweitzer
Forest Service–Southern
 Research Station
U.S. Department of
 Agriculture
Huntsville, Alabama

David Wm. Smith
Department of Forest
 Resources and
 Environmental
 Conservation
Virginia Polytechnic
 Institute and State
 University
Blacksburg, Virginia

Mike Staten
Anderson-Tully Company
Lake Village, Arkansas

Jeffrey W. Stringer
Department of Forestry
University of Kentucky
Lexington, Kentucky

Larry A. Tankersley
Department of Forestry, Wildlife,
  and Fisheries
University of Tennessee
Knoxville, Tennessee

Andrew Vander Yacht
Department of Forestry, Wildlife,
  and Fisheries
Center for Native Grasslands
  Management
University of Tennessee
Knoxville, Tennessee

Tamara Walkingstick
Arkansas Forest Resources Center
University of Arkansas
Fayetteville, Arkansas

Daniel Yaussy
Forest Service–Northern
  Research Station
U.S. Department of Agriculture
Delaware, Ohio

section one

Background and biology:
Setting the stage

chapter one

The mighty oak

Why do we manage oaks?

Bryan Burhans, Patrick D. Keyser,
Craig A. Harper, and Todd Fearer

Although the authors of this chapter grew up in different states at different times, all four of us took for granted the oak trees that surrounded us. As bowhunters, we looked to each fall's mast crop to guide our hunting strategies. We cherish our vivid recollections of early fall days scouring the woods for oaks producing abundant acorns and searching for the best place to hang a deer stand.

Regardless of your background, it is easy to find something exhilarating, enchanting, and satisfying about our oak forests, whether it is cutting firewood, hunting, or simply enjoying their timeless beauty. Years as students, managers, researchers, educators, and leaders in conservation of this great natural resource have added to our perspective. These oak forests are one of the largest and most diverse temperate forest ecosystems on the planet. Scores of species of wildlife call it home. Hunting within them puts millions of pounds of lean, wholesome venison on tables across the country; it also supports a vast recreational economy. And timber harvest provides countless jobs and revenue for our country.

While the lush, canopied forests that blanket the landscape of the Eastern United States—from Southern Wisconsin and New England to Northern Georgia and Central Arkansas—seem enduring, that is not the case. Many of the forces that helped shape today's oak forests have changed, perhaps permanently. Fires are rare and many large herbivores are gone. Deer are at record-setting densities and their natural enemies, the wolf and the mountain lion, no longer keep them in balance with forested habitat. In places, this imbalance is wreaking havoc on the understory vegetation.

Recent data collected by the U.S. Department of Agriculture (USDA) Forest Service's Forest Inventory and Analysis (FIA) project clearly illustrate the dramatic changes that have occurred in our forests and what that portends for the future. These FIA data show that the volume of oaks in the 5.0- to 6.9-inch stem diameter class decreased by more than 35% between 1987 and 2007. These oak saplings are needed to replace the aging oaks in the overstory. Unless something changes, other hardwood species that dominate the understory, such as maples, will eventually replace the oaks when they die or are harvested.

It is important to understand that oaks across much of the Eastern United States do not form the climax state of the forest, but rather they are mid-successional species. In the absence of disturbance, such as a timber harvest or fire, these forests will naturally transition into ones dominated by late-successional species. In Connecticut, for example, the conversion from oak to maple, birch, and beech has occurred at the rate of 5% every decade since 1938. Although some may want to blame overharvesting, the loss of oaks is not occurring in the overstory. The real problem is that we are not producing the young oaks required to replace existing, mature oaks, and oak trees do not live forever.

The disappearance of young oaks is not the only serious problem with eastern oak forests. Some species have disappeared, most notably the American chestnut. The loss of this tree during the early 1900s dramatically shifted species composition in many parts of the region's forests. Oaks filled many of the vacancies left by the loss of chestnut trees. However, new species have appeared—many of them unwelcome—such as the gypsy moth, kudzu, Japanese stiltgrass, and Ailanthus. One of the more recent to appear is the ominously named *Sudden Oak Death*, which is caused by the pathogen *Phytophthora ramorum* and is spreading into the region from the West Coast.

All of these changes to oak forests have critical implications that cover the gamut from lost timber value, increased wildfire vulnerability, and decreased habitat quality for a number of wildlife species. Oaks have traditionally been among the most valuable timber produced in eastern hardwood forests. Oak timber has also held that value more consistently than other species. Replacing oak with less valuable species may result in substantial lost revenue to the landowner, and perhaps even those who process that lumber. With respect to wildlife, the consequences appear to be quite substantial. For instance, oak mast is an important source of nutrition for a number of wildlife species, from the easily overlooked white-footed mouse to ruffed grouse and black bear. In some cases, mast makes the difference in the physical condition of animals entering the winter and their reproductive success the following spring. The structure of oaks appears to be important to some species such as the cerulean warbler—a small migratory songbird that lives in the canopy of tall oaks. Recently listed on an endangered species petition, this warbler prefers nesting and foraging in oaks.

Despite all of these problems, we can still promote the growth of healthy oak forests. The very fact that the oak forests we enjoy today did not develop by accident, but rather as a result of various natural disturbances, suggests that science-based management may be able to help improve forest health. Management may not only be a good option, it may be essential given the many changes in this ecosystem that have altered those natural disturbances, added new species, and removed old ones. Simply letting "nature take its course" will result in "unnatural" forests. Oak forests developed with frequent disturbance, and many management practices are simply disturbance that we control. Using science-based forest management practices, you can secure the future of oak forests on your land. You must move beyond focusing on the mature overstory trees and look down. The next oak forest lies at your feet.

Private forest landowners own or control 80% of oak forests in the Eastern United States. Landowners, therefore, have the ability to improve the health of the vast majority of our eastern oak forests by addressing how they manage stands on their property. But they need to understand the challenges facing today's oak forests—and how to best address those challenges—vary widely. It is our goal to provide landowners with the best and most up-to-date, science-based information to help them meet their goals and ensure healthy oak forests for the next generation.

In addition to the information provided in this book, you should seek appropriate technical assistance from qualified professionals. Forest management is a combination of art and science, and a seasoned forester and wildlife biologist can help you achieve your goals.

It is easy to take the mighty oak for granted. Oaks are a common and dominant component of the forest. But, the future of our oak forests is in question unless appropriate management strategies are used.

chapter two

History of eastern
oak forests

Marc D. Abrams

Contents

Ecology of eastern oak forests prior to European settlement 9
Human impacts on oak forests following European settlement 11
Further suggested reading 16

Temperate hardwood forests dominate much of the Eastern United States, from Central Maine to Northern Minnesota and Eastern Texas to North-central Florida. Within the eastern deciduous biome, oak represents one of the dominant forest types. Approximately 30 oak species occur in the Eastern United States, of which white oak, red oak, black oak, scarlet oak, post oak, and chestnut oak are among the most dominant (Table 2.1). Oaks are an important component of many forest associations in the Eastern United States, including northern hardwood conifer, maple–beech–basswood, mixed mesophytic, oak–hickory, oak–pine, and southern ever-green (see Chapter 3). Indeed, oak trees have dominated forests of the Eastern United States during the last 7000–9000 years. The long-term stability of oak forests is remarkable considering all the ecological problems these species are presently facing. The main objective of this chapter is to discuss the ecological history and long-term stability of oak in the Eastern United States and causes for the recent oak regeneration failure.

Table 2.1 Presettlement oak forest types reconstructed from witness tree data recorded in land surveys during the early European settlement period in various regions of the Eastern United States

Region and forest type	State
Northeast	
White oak–black oak–pine	MA
White pine–white oak–black oak	CT
White oak–black oak–hickory	NY
Pine–white oak–chestnut–black oak	MA
Mid-Atlantic region	
White oak–black oak–chestnut	NJ
Red oak–chestnut	NJ
White oak–black oak–chestnut–hickory	NJ
Beech–sugar maple–basswood–white oak	NY
White oak–red oak	VA
Chestnut–red oak	VA
White oak–red oak	VA
White oak–red oak–chestnut oak–hickory–pine	VA
White oak–chestnut–hickory–pine	WV
White oak–pitch pine	WV
White oak	PA
White oak–white pine–hickory	PA
Chestnut oak–white oak–pitch pine	PA
Black oak–white oak–chestnut–hickory	PA
White oak–chestnut–red maple–white pine	PA
White oak–chestnut oak–white pine–Virginia pine	PA
Southeast	
Longleaf pine–slash pine–turkey oak	FL
White oak–black oak	NC
Oak–chestnut	NC
Chestnut–chestnut oak–red oak	NC
Oak–hickory–pine	GA
Oak–red maple–sweetgum	GA
Midwest and Central Plains	
White oak–bur oak–hickory	MO
White oak	MO
Post oak–shortleaf pine–juniper	MO
White oak–black oak–bur oak	IL
Post oak–hickory	IL
White oak–black oak	IL
	(Continued)

Table 2.1 (Continued) Presettlement oak forest types reconstructed from witness tree data recorded in land surveys during the early European settlement period in various regions of the Eastern United States

Region and forest type	State
White oak–white ash–beech	IL
Blackjack oak–post oak	OK
White oak–hickory	OH
White oak–black oak–hickory	OH
Beech–sugar maple–white oak	OH
White oak–red oak–pin oak	TX
North-Central	
White oak–black oak	MI
Bur oak savannah	WI
White oak–black oak	WI
Bur oak–red oak	MN
Red pine–white oak–white pine	MI

Sources: Adapted from Abrams, M.D., *BioScience*, 42, 346, 1992; Abrams, M.D., *Annales des Sciences Forestieres*, 53, 487, 1996; Abrams, M.D., *BioScience*, 48, 355, 1998.

Ecology of eastern oak forests prior to European settlement

Pollen and charcoal preserved in undisturbed sediment in lakes and bogs can be dated and analyzed, and these data provide an invaluable source of information on the long-term changes of vegetation and fire occurrence. Starting 10,000 years ago, the Eastern United States was dominated by pine forests in the Southeast and mid-Atlantic regions, and spruce forests in the mid-Atlantic, midwest, and central regions. During that time, oak occurred in the South and Southeast, but it represented only 1%–5% of the pollen percentages, compared with 20%–40% for pine. By 9000 to 7000 years ago, oak abundances started to increase dramatically, and this genus has dominated most eastern forests ever since. For example, on the Cumberland Plateau of Tennessee, a warming trend in the region 8000–5000 years ago was reflected in higher pollen influxes for oak, ash, and hickory, and increased the presence of charcoal from aboriginal burning. In addition, about 30 million acres of oak savannah once covered portions of Minnesota, Iowa, Missouri, Illinois, Wisconsin, Michigan, Indiana, and Ohio. Native Americans of the Central Plains used fire management to promote oak savannahs and tallgrass prairie to provide forage for large herbivores, such as bison, antelope, elk, and deer, as well as a dietary resource for themselves (Figure 2.1). The suppression of Indian fires led to the conversion of oak savannahs and tallgrass prairies to closed oak forests.

Figure 2.1 Understory fire in a bur oak (*Quercus macrocarpa*) savannah in Eastern Kansas. (Photo courtesy of M. Abrams.)

Most scientists believe that recurring surface fire was integral to the historical development and long-term persistence of eastern oak forests. In New England, for example, sediment charcoal levels dating to 2500 years ago were lowest in northern hardwood forests, intermediate in oak and pine forests, and highest in pitch pine and oak forests. In the Hudson Highlands of Southeastern New York, oak forests were maintained throughout most of the Holocene as a result of Native American fires. A dramatic increase in oak, hickory, chestnut, and pine in Eastern Kentucky 3000 years ago corresponded to increased fire and Native American activity. In the Southern Blue Ridge of North Carolina, oak, chestnut, pine, and birch were the dominant tree species during the last 4000 years.

Between 10,000 and 6,000 years ago, an expansion of oak and pine and increase in sediment charcoal occurred in the forests of central Appalachia and the New Jersey Coastal Plain. In Eastern Kentucky, where lightning-caused fires are rare, a large increase in oak, pine, hickory, and chestnut beginning 3000 years ago was associated with frequent Indian fires. In Southern New York, increases in charcoal and oak and pine dominance occurred from AD ~800 to 1300. At Crawford Lake in Southern Ontario, Canada, northern hardwood (beech–maple) forests were transformed to white pine–oak forests in response to Iroquois cultivation and burning during the 1400s. The direct dating of fire scars on trees within old-growth forests provides further proof that periodic fire (in the range of every 5–25 years) occurred in Eastern North American oak forests.

At the time of European settlement of the Eastern United States, the distribution of oak was widespread as reported in numerous witness tree studies. Witness trees were recorded on maps as early surveyors walked along section lines and corners and at property corners during the eighteenth and nineteenth centuries throughout the Eastern United States. A compilation of these data can be seen in Table 2.1, showing just how dominant oak trees were in the Eastern United States when Europeans first arrived. Alternatively, many of the species that have replaced oak today, such as red maple, sugar maple, beech, birch, and blackgum, were present in low numbers in the presettlement oak forests.

Human impacts on oak forests following European settlement

The first major event that followed European settlement was land clearing and the cutting of forests for building material and firewood (Figure 2.2). Momentum increased with a rising population and wood demands of the charcoal iron industry; this escalated into the "Great Cutover" during the nineteenth century. The forest was recognized for its importance for industrialization, agriculture, and the basis of material progress. Although iron furnaces in the United States date back to the early 1700s, their peak activity was in the middle nineteenth century. By 1856, there were 560 iron-producing furnaces, 78% of which used wood charcoal. Each furnace required wood from approximately 150 acres of forest per year over a 50-year period. Areas near the furnaces were often cut at a 25–30-year cycle, creating coppice (sprout) forests. Other industrial uses of wood included railroads, mines (props), ship building, and manufacturing, although the largest consumption remained for domestic uses. Most of New England, New York, the Eastern Seaboard, and the Ohio Valley were already logged at least once by the mid-nineteenth century. Approximately, 99% of the original forest was gone by 1920.

Prior to 1850, commercial forestry in the Eastern United States was typically limited to small sawmills in most towns. Subsequent increases in the demand for timber led to the large-scale commercialization of the forest industry by the second half of the nineteenth century, including steam-powered saws and railroad logging. Logging escalated, leading to the height of the "clearcut era" from 1870 to 1920. Not only were the original forests cut, but there was a large loss of forest area to land clearing (agriculture) during the nation-building period. It has been estimated that 164 million acres of U.S. forests were cleared for agriculture by 1860. Between 1850 and 1920, the forested area of the Eastern United States declined from approximately 230 to 147 million acres. The loss of forested land to agricultural clearing ranged from 22% to 76% (averaging 50%) in 12 Eastern states.

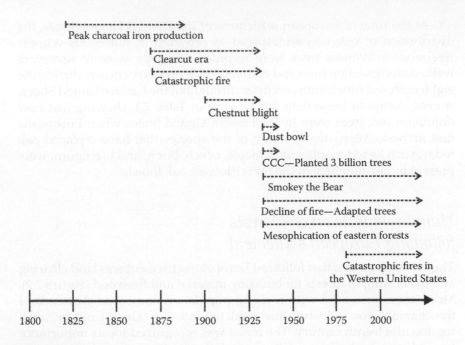

Figure 2.2 Major land use history changes in forests in the Eastern United States since 1800.

Areas left in old-growth forests totaled less than 0.5%. The second-growth forests that formed in the cutover areas were subsequently grazed by livestock (pigs, sheep, and cattle), which resulted in great losses of acorns and other nuts (e.g., hickory, chestnut), tree seedlings, and other understory plants, as well as causing soil compaction and erosion.

The "Great Cutover" logged billions of board feet of timber in the Eastern United States and produced vast areas covered in "slash" (logging debris). As the slash dried, huge wildfires followed, which burned with an intensity not experienced in the original forest (Figure 2.2). These fires include the Great Miramichi and Maine fires in 1825, which burned more than 3 million acres; the Peshtigo fire in Wisconsin and nearby Michigan, which burned 3.8 million acres; the Michigan fire in 1881, which burned 1 million acres; the Wisconsin and Hinckley fires in 1894, each of which burned 1–2 million acres; and the Adirondacks fire in 1903, which burned 0.65 million acres. These fires, coupled with additional huge fires out west (the Great Idaho fire in 1910, which burned 3 million acres), ushered in the fire suppression (Smokey the Bear) era in the United States starting in the 1930s.

What were the impacts of near-complete felling of the eastern forest followed by the catastrophic burning of huge tracts? One might guess "complete devastation," but this would be a misrepresentation. The resilience of eastern forests is one of the great ecological marvels of the world and reinforces the disturbance-tolerant attributes for most of its species. That is not to say that profound changes did not occur; they did, but what came later in the twentieth century was much more devastating. Following the vast clearcutting and subsequent wildfires, the remaining eastern forests came back initially much the same as they were before. When the compositions of presettlement versus more recent (1960–1982) forests in four locations in the Northeastern United States are compared, most major species changed very little. Exceptions to this include an increase in sugar maple and red maple in some areas, a decline in beech and hemlock, and an increase in red oak and chestnut oak. Most of these changes were less than 10%, although the beech decline was larger. A large decline in white pine also took place, which is attributed to it being selectively logged during the early clearcut era and its inability to sprout back like most hardwoods. The Lakes States experienced a large increase in aspen–birch forest as a direct result of clearcutting and burning the original pine and hardwood forests. Another casualty of the clearcut era was white oak, one of the east's most dominant tree species, which declined in many eastern forests between the presettlement and present day. This decline is attributed to white oak not being as well adapted to intensive disturbance regimes compared with northern red oak and chestnut oak. These species increased as a result of white oak's decline, as they did with the decline in chestnut from the blight (see below). The opportunistic nature of red oak is very apparent in central Wisconsin where it increased from 1%–2% to 37%–51% importance in certain areas following the clearcutting and burning of the original forest.

One of the most profound and permanent impacts on the eastern forest during the early twentieth century did not come as a result of clearcutting and burning, but from exotic diseases. The most devastating example of this is the chestnut blight fungus, which was introduced to America in the early 1900s (Figure 2.2). Chestnut was once a dominant species along the Appalachian Mountains from Maine to Georgia, where it typically represented 10%–20% of the forest composition along with oaks and hickories. Sometimes chestnut's dominance rose to 50% of the stems and 90% of the volume. As a result of the blight, nearly 100% of the chestnut trees were extirpated, relegating the species to understory sprouts (the blight does not kill the root system). As devastating as this loss is, the resilience of the eastern forest to disturbance was once again expressed. As the chestnut was killed off, neighboring trees such as oak and hickory filled the chestnut gaps, thereby maintaining productive forests. Red maple, however, also started its rise in the eastern forest at this time in response to the chestnut blight and clearcut era logging.

The ecology of tree species can be dramatically altered by native invasive plant and animal species. This occurs when changes in land use allow certain plants or animals to increase to population levels far beyond those in the presettlement forest. One important example in this category is white-tailed deer, which increased rapidly during and after the clearcut era in many Eastern states. This is attributed to the increased availability of browse near the forest floor coupled with greater hunting restrictions. Deer will browse particularly intensively on hardwood sprouts formed after logging or fire. Many areas, especially in the Northern and Northeastern United States, Southern Canada, and many parts of the Appalachians, have deer-browsing intensities so severe that regeneration failure of most major tree species is now occurring even in trees that are not particularly favored by deer (see Chapter 17). When deer density is high, it is not unusual for forests to contain little or no tree regeneration and have understories populated with fern, grasses, and sedge avoided by deer. I consider the continuation of high deer densities and intensive browsing in the East to be among the greatest risks to forest health that exist today.

The suppression of forest fires since the 1930s, known as the "Smokey the Bear" era, has caused the greatest change in the ecology of eastern oak forests, along with deer browsing (Figure 2.2). Indeed, this topic has received a huge amount of attention in the forestry literature. It is now well recognized that periodic burning is an ecological requirement for the long-term perpetuation and health of most oak, hickory, and northern and southern pine forests. As a result of post-1930s fire suppression, most of these forests have undergone dramatic changes in composition and structure, mostly as a result of successional replacement by more shade-tolerant species. Exceptions to this might include forests on very dry sites that retard the development of the oak replacement species. Fire suppression leads to rapid increases in tree density, moving these systems to closed canopy forests where oaks (and others) can no longer regenerate. The main oak replacement species is the shade-tolerant red maple, whose range covers nearly the entire eastern forest (Figure 2.3). Other tree species increasing in eastern oak forests include sugar maple, beech, hemlock, blackgum, white ash, black birch, tulip poplar, and black cherry.

It is interesting to note that nearly all late-successional trees, with the exception of blackgum, are sensitive to fire and are readily killed by repeated burning. The range of beech and blackgum, like red maple, includes nearly the entire eastern forest. The range of sugar maple covers the entire northern half of the eastern forest, whereas hemlock's range covers the Northern states and the entire Appalachian chain. The scarcity of fire-sensitive, late-successional tree species in the pre-European forest (outside of the northern hardwoods; Table 2.1) is probably best explained by the role of periodic fire by Native Americans. An analysis of this phenomenon indicated that tree species showing the largest decrease from

Figure 2.3 High density of pole-size red maple in a white oak black oak forest in central Pennsylvania that has not been burned since the 1930s. (Photo courtesy of M. Abrams.)

presettlement to present day are the fire-dependent oak, chestnut, hickory, and pine. The increase in northern red oak, black oak, and chestnut oak in some present-day forests appears to be temporary, as they have had very little recruitment in the last half-century.

In conclusion, extensive logging and land clearing; wildfires; the introduction of exotic insects, diseases, and invasive plants; increased deer browsing; and the Smokey the Bear era have led to unprecedented and rapid changes in oak forest composition and structure. The Eastern United States has seen the conversion of about 50% of its forests and 99% of its tallgrass prairie to agriculture; the extirpation of the once-dominant chestnut, elm, and butternut species from exotic pathogens; the loss of many white pine forests; the decline in the super-abundant white oak; and the virtual cessation of oak, hickory, and pine canopy recruitment and the rise of red maple (and others) in its remaining forests. The replacement of fire-adapted oak and hickory trees by later successional species has resulted in a "mesophication" (cooling, dampening, and shading) of the forest understory, rendering these ecosystems less prone to prescribed or natural fire and less conducive to subsequent oak and hickory recruitment (Figure 2.4). Indeed, the eastern landscape has undergone a near-complete transformation over the last 350 years. Anthropogenic impacts on forests during the late nineteenth and twentieth centuries have even been characterized as a "perfect storm" (Figure 2.2). I have argued that the impacts of Indian land use practices in the original eastern forest

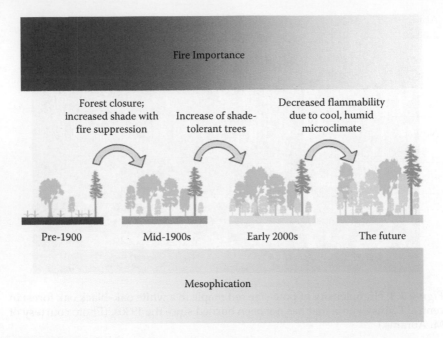

Figure 2.4 The mesophication of eastern oak forests as a result of fire suppression. (Adapted from Nowacki, G.J. and Abrams, M.D., *BioScience*, 58, 123, 2008.)

and tallgrass prairie were ubiquitous, not merely localized. Most forests and grasslands were actively managed to meet the dietary needs of the Native Americans. But the factors that perpetuated vast upland oak forests and savannahs and tallgrass prairie during the last 7000+ years were profoundly altered in the few centuries following European settlement. This change in land use history (especially fire suppression) and the large increases in deer populations are largely responsible for the lack of oak regeneration and recruitment into the canopy presently occurring in eastern forests.

Further suggested reading

Abrams, M.D. 1992. Fire and the development of oak forests. *BioScience* 42: 346–353.

Abrams, M.D. 1996. Distribution, historical development and ecophysiological attributes of oak species in the eastern United States. *Annales des Sciences Forestieres* 53: 487–512.

Abrams, M.D. 1998. The red maple paradox. *BioScience* 40. 355 361.

Abrams, M.D. 2002. The postglacial history of oak forests in eastern North America, in W. J. McShea and W. M. Healy (eds), *Oak Forest Ecosystems: Ecology and Management for Wildlife*. John Hopkins University Press, Baltimore, MD, pp. 34–45.

Abrams, M.D. 2003. Where has all the white oak gone? *BioScience* 53: 927–939.

Abrams, M.D. and Nowacki, G.J. 2008. Native Americans as active and passive promoters of mast and fruit trees in the eastern USA. *The Holocene* 18: 1223–1137.

Delcourt, P.A. and Delcourt, H.R. 2004. *Prehistoric Native Americans and Ecological Change*. Cambridge University Press, Cambridge, U.K., 203 pp.

Lorimer, C.G. 2001. Historical and ecological roles of disturbance in eastern North American forests: 9,000 years of change. *Wildlife Society Bulletin* 29: 425–439.

Nowacki, G.J. and Abrams, M.D. 2008. The demise of fire and "mesophication" of forests in the eastern United States. *BioScience* 58: 123–138.

Pyne, S.J. 1982. Fire in America: A cultural history of wildland and rural fire. Princeton University Press, Princeton, NJ, 654 pp.

Whitney G.G. 1994. From coastal wilderness to fruited plain: A history of environmental change in temperate North America from 1500 to the present. Cambridge University Press, New York, 451 pp.

chapter three

Description and distribution of oak forests across the Eastern United States*

Christopher M. Oswalt and Matthew Olson

Contents

Introduction 19
   From global to local perspective 19
   Site quality: Influence on distribution 20
   Methods of assessing site quality 20
Geographic distribution of oaks 22
Major regions 22
   The Northern Hardwood Region 22
   The Central Hardwood Region 25
      The Ohio Valley 27
      The Allegheny Hardwoods 27
      The Appalachian Hardwoods 29
      The Ozark/Xeric Upland Hardwoods 30
   The Southern Hardwood Region 32
      The Bottomland Hardwoods 33
      The Coastal Plain Oak–Pine 36
   The Oak Savannahs and Prairie Region 38
Further suggested reading 40

Introduction

From global to local perspective

Oaks are found across the globe. In fact, approximately 600 species have been identified worldwide. Native to the Northern Hemisphere, oaks are found from the cold latitudes of North America to the tropics in Asia and

---

* The authors of this chapter relied on the available scientific literature, but for ease of reading, we have not cited references in the body of the text. Credits for that information are given in the "Further Suggested Reading" section at the end this chapter.

the Americas. In North America, oaks are widely distributed and found in both the Western and Eastern United States with only two species (chinkapin and bur oak) common to both regions. Oak species richness (the number of oak species) in the Eastern United States is highest in the South near the shared borders of Alabama, Florida, and Georgia and declines as you travel north and west. At this point, the ranges of many oak species, both common and not so common, overlap. Due to the sheer number of oak species found in the United States, a variety of management strategies are required because each species has individual requirements for proper management. However, an understanding of the influence of site quality or site characteristics helps landowners and managers to manage the diversity of requirements that exist. Environmental conditions not only help define a forested site but will often dictate the species that can be established and flourish. Therefore, an understanding of site conditions equates to an increased probability of successful oak management strategies.

Site quality: Influence on distribution

Managing oak forests is both challenging and rewarding. Each of the common oak forest systems in the Eastern United States offers unique challenges dictated by local species composition and site quality. The distribution and prominence of any one oak species are highly influenced by climate and landform. Locally, whether oaks are found at a particular site or not is influenced by soil moisture, geology, and physiography. Site differences such as slope, moisture, aspect toward the sun, and soil characteristics all can influence the composition of each oak forest and the manner in which you approach its management. Those differences may be obvious, as in the Appalachians, or very subtle, as is the case in bottomland oak forests. However, these site characteristics will not only determine the oak species growing in your forest today, but also the ease or difficulty of managing for oaks in the future.

Methods of assessing site quality

As you walk in your woods, keep in mind that the success or failure of your oak management plan will largely depend on correctly evaluating your forest site. Look around; take note of what is growing, both in the upper canopy and beneath your feet. Examine the soil. Is it deep, rocky, or largely sand or clay? Is there standing water or are the soils generally wet? If there is a prominent slope, what direction is it primarily facing? Answers to each of these questions will help guide you through many decisions you will be making about how best to manage your oak forest in order for it to remain an oak forest. Understanding your site is the key to making wise management decisions.

WHAT IS A "SITE?"

A forest site can be defined as a contiguous area sufficiently uniform in quality and with similar soils, landform, and vegetation, generally similar enough to be distinguishable as a unit. In very general terms, a forest site should be viewed as a unit that when treated (harvested, planted, etc.), there is a reasonable expectation that the biological response will be similar across the space.

Characteristics of good sites:

- Well-drained stream and river bottoms
- Mountain coves
- Mid and lower slopes facing north and east
- Lower slopes facing north or southeast
- Gradual topography
- Deep, well-drained soils

Characteristics of poor sites:

- Upland ridges
- Upper and mid-slopes facing south, southwest, and west
- Steep slopes
- Soil poorly drained
- Soil excessively drained, droughty
- Soils high in clay or sand
- Soils are largely rock or gravel

The best oak sites are usually on medium-textured soils (loamy soils) because nutrient and moisture availability are adequate. Coarse-textured soils (sandy soils) have limited moisture-holding capacity and result in a lower quality site. Conversely, fine-textured soils (clayey soils) can often hinder site drainage and result in too much moisture. Some of the best oak sites are also north- and east-facing. South- and west-facing slopes receive more direct sunlight during the growing season, which can limit moisture availability (Table 3.1).

In the remainder of this chapter, we will guide you through some of the common oak forest types in the Eastern United States and the sites where they are found. As you read this section, look for similarities with your oak forest. Imagine yourself there and make comparisons between your forest and what you read here. Of course, each forest is unique, but there are many similarities that, when acknowledged, will help you manage your property to reach its potential and meet your objectives.

Table 3.1 Common oak species and general characteristics of sites where each is commonly found

Common name	Site characteristics
Black oak	Uplands
Cherrybark oak	Floodplains
Chestnut oak	Dry ridges; xeric soils
Northern red oak	Moist slopes; well-drained uplands
Overcup oak	Lowlands and floodplains
Post oak	Uplands; sandy or dry clay soils
Shumard oak	Moist soils near streams or swamps
Southern red oak	Dry, sandy uplands; heavy clay soils
Water oak	Moist uplands; lowlands and floodplains
White oak	Well-drained soils

Source: Adapted from Ober, H.K. and Minogue, P.J., Managing oaks to produce food for wildlife. Wildlife Ecology and Conservation Department, Florida Cooperative Extension Service, Institute of Food and Agricultural Sciences, University of Florida, Publication WEC249, 2008.

Geographic distribution of oaks

Major regions and subregions: The eastern half of the United States is approximately one-half forested. In the Eastern United States, there are four primary oak-growing regions (Figure 3.1): (1) the Northern Hardwood Region, (2) the Central Hardwood Region, (3) the Southern Hardwood Region, and (4) the Oak Savannahs and Prairie Region. With the exception of the Oak Savannahs and Prairie Region, we have identified at least two subregions within each region. Each region and subregion will be explored in order to offer a brief description that may be helpful to compare to your particular forest, which may aid in fine-tuning your management options.

Major regions

The Northern Hardwood Region

The heavily forested region straddling the border between the United States and Canada, extending from Maine to Minnesota, is called the *Northern Hardwood Region*. In fact, approximately 75% of the land area within this region is forested, and in Maine, the nation's most heavily forested state, 90% of the land base is forested. In the United States, the Northern Hardwood Region occurs in more than 10 states. Some ecologists consider this region transitional between the Central Hardwood Region to the South and the Boreal Forest Region to the North, because the geographic ranges of several tree species common to both regions

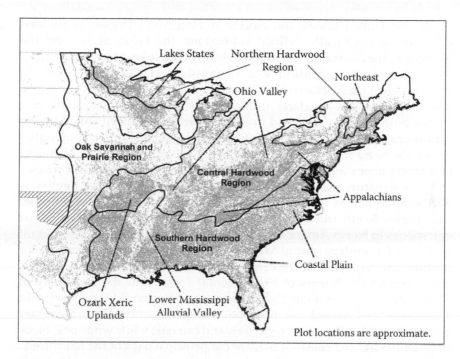

Figure 3.1 Major oak-growing regions of the Eastern United States and occurrence of oak. Dots indicate locations where oak was sampled in the most recent surveys (approximately 2002–2007) of the USDA Forest Service Forest Inventory and Analysis Program. Regional boundaries follow Bailey (1997). No current data were available for Western Oklahoma. (Adapted from Johnson, P.S. et al., *The Ecology and Silviculture of Oaks*, CABI Publishing, New York, 2002, 544 pp.)

overlap within the Northern Hardwood Region. For example, it is not uncommon for oak, hickory, or maple species of the South to form mixed-species stands with aspen, spruce, or balsam fir species, all of which have a more northern affinity.

It is the Northern Hardwood Region that is so well known for its scenic displays of leaf coloration during the fall, which attracts "leaf lookers," or people living outside the region who come to view the fall foliage, each autumn. The signature tree species of the Northern Hardwood Region is sugar maple, which is the same species maple syrup is produced from and which is featured prominently on the Canadian flag. Oaks, on the other hand, are largely a minor component of the Northern Hardwood Region. Although their abundance is low region-wide, oaks generally become more abundant within southern portions of the Northern Hardwood Region. This increase in oak abundance is associated with the shift from cooler and wetter to warmer and dryer conditions moving from north to south.

In the United States, the Northern Hardwood Region is divided into two geographically defined subregions: the Lakes States and the Northeast. The Northern Hardwood Region of the Lakes States falls within the northern sections of Minnesota, Wisconsin, and Michigan. The northeast portion of the Northern Hardwood Region covers most of the Central and Northern New England (Maine, New Hampshire, Vermont, and Massachusetts), New York, and much of Northern Pennsylvania. Moving southward, forest types of the Northern Hardwood Region grow at higher elevations, as far south as the Appalachians of North Carolina, north- and east-facing slopes, and on cooler landforms like shady valleys and coves.

The signature oak species of the Northern Hardwood Region is northern red oak as it tends to grow on cooler, moister sites than other oak species. Northern red oak is also one of the most widely distributed oak species in North America. This oak can be a dominant or codominant species as far north as central Maine. For example, it is not uncommon for northern red oak to share the canopy with eastern white pine, red maple, and aspen on the Penobscot Experimental Forest near the University of Maine, which is more than 200 miles north of Boston, Massachusetts. In addition to northern red oak, principal oak species of the Lakes States are white oak, black oak, northern pin oak, and bur oak, while white oak, black oak, scarlet oak, and chestnut oak are the principal oaks of the Northeast.

Topography, soils, and disturbance history play an important role in structuring the distribution of oak species within landscapes of the Northern Hardwood Region. Oaks are often found on warmer, drier sites where they are more competitive than mesophytic species (a name given to species that tend to occur on moist—or mesic—sites) such as maples, ashes, and birches. Mesophytic species have a decidedly competitive advantage over oaks on cooler, moister sites. Where topography is rolling or hilly, oaks tend to occur on south-facing slopes, because the greater exposure to incoming sunlight makes these sites dryer than other exposures. On landscapes of higher, flat terrain dissected by deep, stream valleys, oaks tend to dominate the higher ground between streams. In mountainous areas of the Northern Hardwood Region, oaks are often found growing on broad, valley-floor sites and south-facing slopes at mid-elevations. Soils formed of coarse material (sand) tend to be dry since they hold less moisture against the pull of gravity and, therefore, tend to support oak. In the Lakes States, northern pin oak, white oak, and black oak persist on xeric (dry), nutrient-poor soils formed from deep sands. Similar sand plains are common to parts of the Northeast, which also support oak stands. Interestingly, these sandy deposits were formed thousands of years ago by meltwater from retreating glaciers.

Disturbance is another important factor controlling the distribution of oak species in the Northern Hardwood Region. Oak-dominated stands are often found on sites with a history of logging and fire. In some cases, the effects of these disturbances can override the influences of topography and soils, enabling oaks to grow on sites where mesophytic species would ordinarily dominate.

The Central Hardwood Region

The East-Central portion of the United States, which is primarily composed of deciduous broadleaf tree species with very few commercially important conifer species, is known as the *Central Hardwood Region*. Covering approximately 220 million acres, the Central Hardwood Region extends from Cape Cod in Massachusetts to Eastern Oklahoma and as far south as Tennessee. One-half of the region is forested and primarily in private ownership within small (less than 50 acres) holdings. Oak–hickory forests dominate the region (Figure 3.2).

The use of fire by Native Americans within the Central Hardwood Region helped shape both the extent and composition of the area's presettlement forests, which influence where oak is found today. Moreover, the importance of oak within the Central Hardwood Region was increased by the demise of the American chestnut in the early 1900s. Today, oak

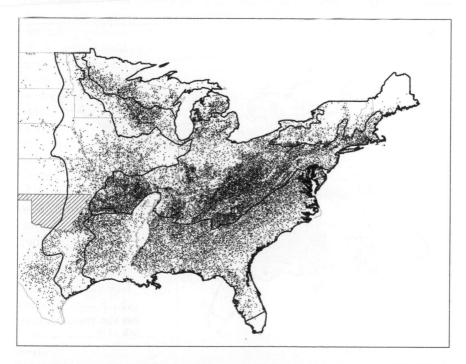

Figure 3.2 Distribution of the Oak/Hickory forest type group as sampled by the USDA Forest Service Forest Inventory and Analysis Program.

forests in the region often develop on sites that are relatively dry. Oaks can easily persist as dominant individuals within these forests, facilitated by their drought tolerance and by sufficient light penetrating a more open canopy and reaching reproduction on the forest floor. Oak species share dominance with other non-oaks in the more mesophytic forests of the region. These forests are generally more complex, and in the absence of a disturbance such as fire, oaks are often displaced in succession by species that are more shade tolerant. In addition, these highly productive systems often lack oak reproduction below very dense overstories.

Predominant oaks in the region include black, white, scarlet, chestnut, post, northern red, and southern red. White oak, for example, is widely distributed across the region with the highest levels of basal area in the eastern half (Figure 3.3). Forests within the Central Hardwood Region are extremely diverse, containing numerous oak and non-oak species. Between 2003 and 2007, the USDA Forest Service Forest Inventory and Analysis (FIA) program, responsible for inventorying and monitoring all U.S. forests, sampled 112 different tree species (22 different oak species) in Tennessee forests classified as belonging to the oak/hickory forest type group. Common non-oak associates within the region are hickories, sassafras, flowering dogwood, blackgum, black cherry, and red maple.

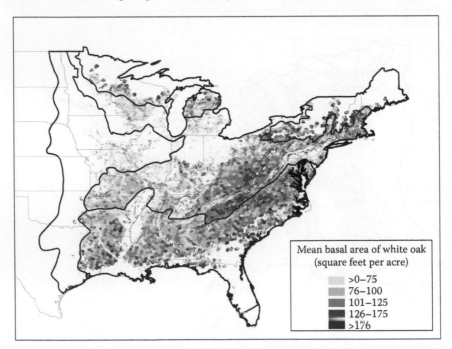

Figure 3.3 Mean basal area of white oak for the Eastern United States as sampled by the USDA Forest Service Forest Inventory and Analysis Program.

The Central Hardwood Region can be subdivided into four separate subregions: (1) the Ohio Valley, (2) the Allegheny Hardwoods, (3) the Appalachian Hardwoods, and (4) the Ozark/Xeric Upland Hardwoods. These four subregions create a region rich in topographic variety and variable in subregional climate. Elevation within the Central Hardwood Region can be as high as 6000 feet in the Appalachians, 1000–3000 feet in the Cumberland Mountains, or "rolling" as in the western half of the region. The mean annual temperatures range from 40°F to 60°F, while the mean annual precipitation ranges from as low as 20 inches in the Ozark/Xeric Uplands, to 65 inches in the Southeast, to 80 inches in the Appalachians.

The Ohio Valley

The Ohio Valley extends from Southwestern Pennsylvania along the Ohio River southwest through Ohio, Indiana, Illinois, and Kentucky to the border shared by Illinois, Kentucky, and Missouri. The area has extensive forest cover and can be characterized as having fairly rugged topography in the eastern portion, to rolling and flat in the West near the confluence of the Ohio and Mississippi Rivers.

According to recent (2008) FIA data, the white oak/red oak/hickory (Figure 3.4) is the most common forest type in the Ohio Valley. In addition, the white oak forest type and the yellow poplar/white oak/northern red oak forest types are common oak forest types in the area (Figure 3.5). Forests in this area, while generally dominated by oak species, contain a significant number of tree species that help create complex associations. The overstory is generally dense and well developed. Similarly, midstories are typically well developed and may be dominated by shade-tolerant species such as flowering dogwood or red maple. Red maple is fairly ubiquitous across the Ohio Valley (Figure 3.6) and is generally a component of every forest in the area. Along with white oak, yellow poplar and white ash are very important species in these forests, particularly found on moist sites, and are widely distributed across the Ohio Valley. For the most part, sites in the area are considered moderate to high quality. For this reason, in the absence of some types of disturbances, natural or man-made, many of the current oak forests of the area may develop into other types dominated by shade-tolerant species.

The Allegheny Hardwoods

The Allegheny Hardwoods describe the diverse, hardwood-dominated forest extending southward from Northwestern Pennsylvania through Western West Virginia down to Eastern Kentucky. The term *Allegheny* refers to the northern section of the Appalachian Plateau, which is called the Cumberland Plateau from Southeastern Kentucky through Eastern Tennessee south to Northern Alabama. The topography of the Allegheny Plateau varies greatly, and relief (change in elevation from the valley floor to the plateau top) may range up to 2000 feet in some places.

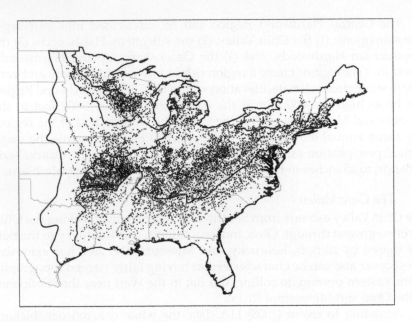

Figure 3.4 Distribution of the white oak/red oak/hickory forest type as sampled by the USDA Forest Service Forest Inventory and Analysis Program.

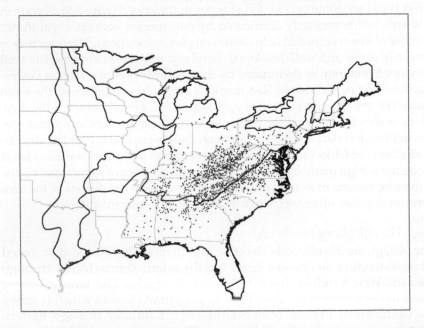

Figure 3.5 Distribution of the yellow poplar/white oak/northern red oak forest type as sampled by the USDA Forest Service Forest Inventory and Analysis Program.

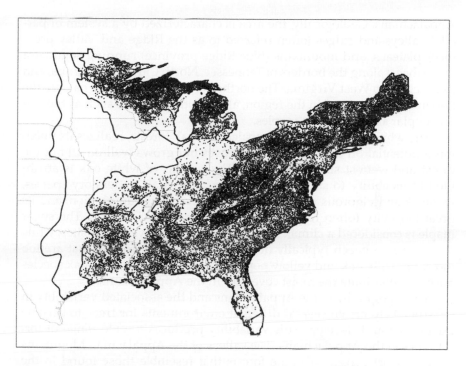

Figure 3.6 Distribution of red maple as sampled by the USDA Forest Service Forest Inventory and Analysis Program.

The Allegheny Hardwoods are sometimes considered a Southern variant of the Northern Hardwoods Region because sugar maple is an important component. However, oaks occur in greater abundance in the Allegheny Hardwoods than in the Northern Hardwood forest. Northern red oak is one of the most important oak species of the Allegheny Hardwoods, which reflects the more mesic nature of the Allegheny Hardwood forest. The dryer soils that form on the plateau tops often support a greater abundance of oaks where a mixture of white oak, black oak, scarlet oak, chestnut oak, and northern red oak can coexist within the same stand. As with the Northern Hardwood Region, higher-quality sites generally support the mesophytic, non-oak species, including sugar maple, black birch, black cherry, and white ash. However, oaks may occasionally dominate high-quality sites with a history of logging and fire.

The Appalachian Hardwoods
The Appalachians are one of the most biologically diverse locations in the world, outside of the tropics. Many of the tree species represented throughout the Central Hardwood Region are also found in the

Appalachians. Geologically, the area is characterized by a system of parallel valleys and ridges (often referred to as the Ridge and Valley province), plateaus, and mountains (Blue Ridge province) running southwest to northeast along the borders of Tennessee, North Carolina, Virginia, and Kentucky into West Virginia. The northern red oak and white oak are the two major oak species in the region, along with non-oaks such as the yellow poplar and numerous maples.

The white oak, within the Appalachian oak forests, realizes its maximum potential on deep, rich soils of coves but grows well on all but the driest and wettest sites in the area. The success of white oak is attributed to its ability to survive for long periods as an understory species, its quick and vigorous response to release from this suppression, and its great longevity (often reaching 400–600 years) (see FORSite). The sugar maple is considered a climax species on some of the very best sites, while the American beech typically dominates many cove forests that are too moist for white oak and yellow poplar. For the most part, non-oak species normally dominate the moist cove sites in the Appalachians.

The topography of the Appalachians and the associated variability in substrate help create several different environments for trees to become established and develop. This variability produces a wide range of forest types in the Appalachians. The valleys of the Appalachian Mountains often support mixed oak–pine forests that resemble those found in the neighboring Piedmont of the Southern Hardwood Region. Look for black, scarlet, chestnut, and white oaks as well as shagbark and pignut hickories. Virginia pine, shortleaf pine, and sassafras may also be present. In addition, the drier and warmer west- and south-facing slopes at lower and mid-elevations often support open oak and southern pine forests. For example, you may find mid-elevation communities comprising pure stands of table mountain pine. The north- and east-facing slopes, in addition to higher elevations, can support forest types similar to those found in the Northern Hardwood Region or even spruce-fir forests at the highest elevations. However, xeric (dry) southern pine stands extend up to exposed ridges, while moisture-loving hemlock and Northern Hardwood forests can reach into protected coves and ravines where temperatures are cooler and moisture is higher.

The Ozark/Xeric Upland Hardwoods

The Ozark Uplands (sometimes referred to as the Ozark Highlands, a smaller component of the Ozark/Xeric Uplands) comprise one of the largest contiguous areas dominated by oak hickory forests within the Central Hardwood Region. Many of these forests probably originated when oak savannahs became closed-canopy forests following fire suppression and settlement beginning in the mid-1800s. The Ozark/Xeric Uplands are located in Southern Missouri and Northern Arkansas and extend into

Northeastern Oklahoma. The area of the Ozark/Xeric Uplands is the most extensive mountainous region between the Appalachians and the Rocky Mountains.

White oak, black oak, and scarlet oak dominate many of the forests in this area. Sites within the Ozark/Xeric Uplands are typically drier and less productive than other areas in the Central Hardwood Region. Therefore, many moisture-loving tree species are not represented. For example, the yellow poplar/white oak/northern red oak forest type is scarcely found in the area (Figure 3.5). Tree species diversity is approximately half that of the Appalachian Hardwood subregion. The driest upland oak sites, typically with very shallow, rocky soils, include species such as post oak, blackjack oak, and at times a mixture of oaks and shortleaf pine, particularly in areas with sandstone bedrock. The post oak/blackjack oak forest type, a very drought-tolerant oak forest type, is common in uplands of this area (Figure 3.7). Sites adjacent to streams (riparian areas) may include river birch and silver maple. Many oak forests on public lands in this area are now being managed to restore oak savannahs and woodlands, areas with open, low-density oak trees, typical of what predominated in presettlement times.

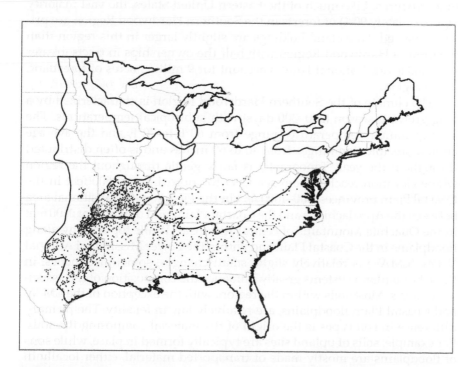

Figure 3.7 Distribution of the post oak/blackjack oak forest type as sampled by the USDA Forest Service Forest Inventory and Analysis Program.

Unlike many oak forests in the Eastern United States, oak reproduction is not a limiting factor when managing for future oak stands in the Ozark/Xeric Uplands. Many oak forests of the area have a sparse midstory with considerable sunlight filtering through open canopies and reaching the forest floor. The sufficient sunlight for reproduction, coupled with few competing drought-tolerant tree species, allows for oak reproduction to establish and develop into large size classes. Large oak reproduction is needed to vigorously respond to natural disturbance and harvesting in order to facilitate the development of the next cohort of oak trees.

The Southern Hardwood Region

The Southern Hardwood Region, sometimes referred to as the Southern Pine-Hardwood Region, stretches from Eastern Texas to Southern Florida to the Atlantic Coast of Georgia, South Carolina, North Carolina, Virginia, and Maryland. The region covers approximately 270 million acres of which roughly 60% is forested. Forests in the region include broadleaved (primarily oak dominated), conifer forests, and numerous hardwood-pine mixtures. Like much of the Eastern United States, the vast majority (approximately 90%) of forests in the Southern Hardwood Region are privately owned. Forestland holdings are slightly larger in this region than the Central Hardwood Region with half the ownerships in tracts greater than 100 acres. National Forests account for 9 million acres of forestland in the region.

The climate of the Southern Hardwood Region is characterized by a long growing season (200–300 days) with subtropical temperatures. The average annual temperatures range from 60°F to 70°F, and the average annual precipitation ranges from 40 to 60 inches and is often distributed throughout the year. Topography is fairly gentle throughout the region where elevation reaches from sea level to approximately 600 feet in the Coastal Plain provinces, 300–1000 feet in the Piedmont (a transitional area between the Appalachians and the Coastal Plain), and upwards of 2600 feet in the Ouachita Mountains of central Arkansas. While topography along floodplains in the Coastal Plain and within the Lower Mississippi Alluvial Valley (LMAV) is relatively slight, small changes, sometimes inches, in these bottomland systems greatly influence the composition of the associated forests. Most soils within the region, with the exception of the LMAV and Coastal Plain floodplains, are relatively low in fertility. The primary difference in soil types is the origin of the material composing the soils. For example, soils of upland sites are typically formed in place, while soils of floodplains are mostly made of transported material, either locally in the case of minor stream systems, or from great distances, as is the case for the floodplain of the Mississippi River.

While this region contains a significant amount of upland oak and upland oak–pine forest types, numerous bottomland oak forests are found within the LMAV and intimately intermingled within the upland forests of the remainder of the Southern Hardwood Region. In reality, the two very different forests (upland vs. bottomland) could be treated as separate regions; however, the tight intermingling of the two types and the somewhat linear nature of bottomland forests associated with riverine systems make it impractical to do so. Therefore, the two are both described as separate site types located within the Southern Hardwood Region: bottomland hardwoods and Coastal Plain (upland) oak–pine. Bottomland hardwood forests cover approximately 27 million acres, while upland hardwoods (dominated by oak–pine) cover the reminder of the forested area within the region. Lack of good-sized oak reproduction is of major concern throughout the region.

Oaks that can be found on upland oak–pine sites in the region are black, blackjack, chestnut, post, scarlet, and southern red, among others with various levels of a pine component. Southern bottomland hardwoods typically include 11 species of oak: cherrybark, delta post, laurel, Nuttall, overcup, pin, Shumard, swamp chestnut, water, white, and willow oaks.

The Bottomland hardwoods

Bottomland forests are distinctly different from their upland counterparts in the region. These forests differ in species composition, ecology, and silvicultural practices, particularly those practices designed to promote regeneration of oak forests. The high level of productivity of bottomland hardwood systems only increases the complexity and also the potential for reward of managing these forests. Numerous bottomland oaks, cherrybark oak (Figure 3.8) for example, are highly prized for their great production and form.

Flooding is relatively common on most floodplain or bottomland sites, and each can differ according to the duration, timing, depth, and frequency of flooding events. Each characteristic acts as an environmental sieve that defines the composition and structure of a bottomland forest. For instance, cherrybark oak (Figure 3.8b) occurs across the region on well-drained terrace sites, while overcup oak (Figure 3.8d), which is very tolerant to standing water, occurs on many of the wetter sites across the region. The duration or time for which water is on a given site can be days, such as small streams of the Atlantic Coastal Plain, or months as can be the case for moderately sized tributaries of the Mississippi River. The timing of a flood can be quite influential if flooding occurs during the growing season, particularly if water levels are deep enough to overtop newly germinating reproduction. Similarly, the frequency in which a site is flooded will dictate which tree species occur because some species are

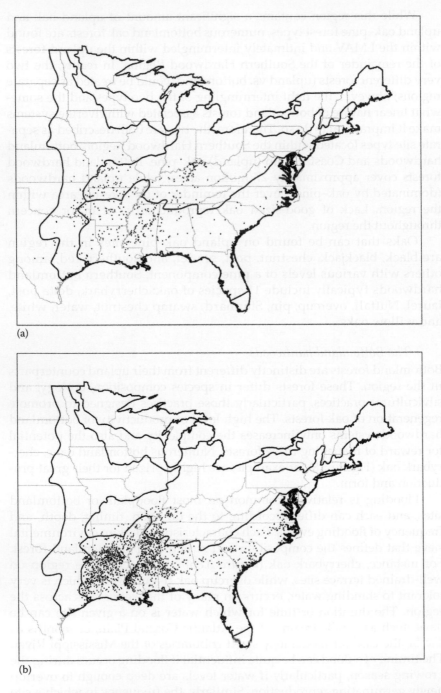

(a)

(b)

Figure 3.8 Distribution of (a) Shumard's oak, (b) cherrybark oak.   (*Continued*)

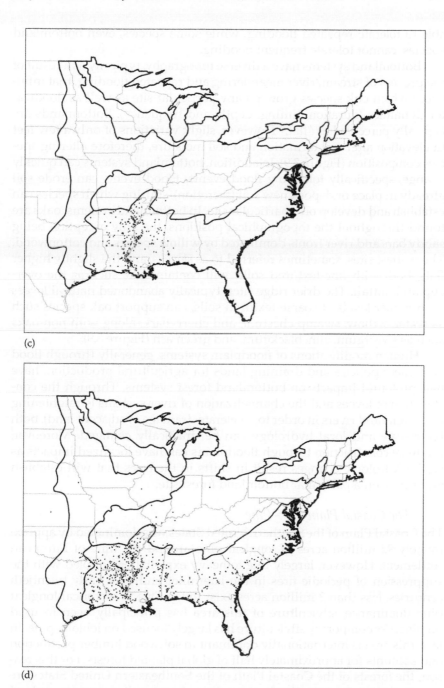

(c)

(d)

Figure 3.8 (Continued) Distribution of (c) Nuttall oak, and (d) overcup oak as sampled by the USDA Forest Service Forest Inventory and Analysis Program.

able to tolerate repeated flooding, while some species, even bottomland species, cannot tolerate frequent flooding.

Bottomland systems have a diverse topography, typically as a result of the process of stream/river meandering and repeated flooding, that influences which oak species grow on any particular site as well as floodwater dynamics (duration, timing, depth, and frequency). Bottomlands are typically perceived as "flat." However, slight variations of only a few feet in elevation alter soil formation and soil moisture, therefore altering species composition (Figure 3.9). In addition, bottomland systems can quickly change, specifically following flood events. Flood events can erode soil already in place or deposit new soil (accretion), shifting which species can establish and develop on a particular site. In bottomland systems, oaks are found throughout the topographical positions, the only exception being sandy bars and river fronts dominated by willow species and cottonwood. The wettest sites, sometimes referred to as *sloughs,* and the slightly higher flats, both with fine-textured soils, will contain oaks such as the overcup and nuttall. The drier ridge sites, typically abandoned natural levees with higher levels of coarse-textured soils, can support oak species such as water, willow, swamp chestnut, and cherrybark, along with non-oaks such as sweetgum, elm, blackgum, and green ash (Figure 3.9).

Human modifications of floodplain systems, generally through flood abatement policies and draining lands for agricultural production, have had profound impacts on bottomland forest systems. Through the construction of levees and the channelization of river systems (straightening of streams and rivers in order to accelerate drainage of adjacent land), both large-scale and local hydrology can be drastically changed. Moreover, roadway construction through floodplains can have localized impacts as well. Hydrologic changes result in shifts in the trees that will establish and thrive on a particular bottomland forest site.

The Coastal Plain Oak–Pine

The Coastal Plain of the Southern United States was dominated by approximately 92 million acres of longleaf pine around the time of European settlement. However, largely due to heavy exploitation coupled with the suppression of periodic fires in the late nineteenth and early twentieth centuries, less than 3 million acres exist today. Due to historical longleaf pine dominance, silviculture of this area has principally been focused on pine. Contemporary silviculture has largely focused on loblolly pine. In fact, this area is internationally dominant in softwood lumber production and accounts for approximately half of global planted forests. For this reason, the forests of the Coastal Plain of the Southeastern United States are among the most productive in the nation. As a result, numerous upland oak and mixed oak–pine forests have been converted to pure pine stands for optimizing per unit area fiber production. However, a large number

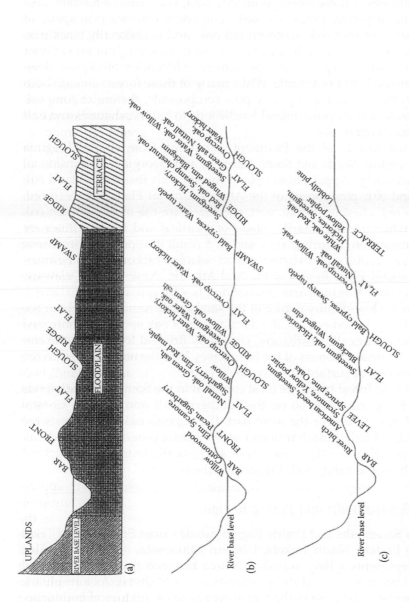

Figure 3.9 Generalized cross sections of major and minor stream bottoms of the Coastal Plain. (a) Major stream valley showing topographic variations. (b) Species associated with topographic variations within a major stream valley. (c) Topographic variations and associated species in a minor stream valley. After Hodges and Switzer (1979). From Hodges (1997).

of upland oak and upland oak–pine forests still exist in the region. The dominant trees of these forests primarily comprise drought-tolerant oaks and southern yellow pines. The oak component consists principally of scarlet oak, chestnut oak, southern red oak, and occasionally black oak. Common pines can include loblolly, shortleaf, and Virginia pine. Other common hardwood species include numerous hickories, blackgum, flowering dogwood, and red maple. While many of these forests contain both upland oaks and southern yellow pine components, the entire compositional gradient from pure upland hardwoods to half hardwoods and half pine to pure pine exists.

Natural forests of the Piedmont, extending from central Virginia through central North and South Carolina and Georgia, are transitional types that bridge Appalachian upland forests with the xeric upland oak forests and oak–pine forests of the Atlantic Coastal Plain. Southern red, northern red, chestnut, white, post, and black are the most common oak species, while pines primarily consist of shortleaf and loblolly. Pines are more common on disturbed sites, and as a result of repeated disturbance over the past century, pines have replaced oaks and hickories in some areas.

In Coastal Plain forests of the Mid-Atlantic (New Jersey, Delaware, Maryland, and Virginia), oak distribution is widespread. This pattern is linked to the low-quality nature of the sandy soils found throughout the Coastal Plain. In some areas, the distribution of oak species is controlled largely by historical disturbances, especially fire and logging. For example, oak-dominated forests of the New Jersey Pine Barrens form when the return interval of surface fire exceeds 40 years.

Oak–pine forest types are most common in the Southern Hardwoods Region, but particularly so on the Southern Gulf and Atlantic Coastal Plains (Figure 3.10). For the most part, the region's oak–pine forests are commonly loblolly pine/hardwood and Virginia pine/southern red oak forest types (Figure 3.11). These forests are typically found on dry, upland sites with coarse, sandy soils throughout the region.

The Oak Savannahs and Prairie Region

The Oak Savannahs and Prairie Region extends from Southeastern Texas north to Eastern North Dakota, Western Minnesota, and Canada. This region represents a large transitional area between the heavily forested Eastern United States and the grasslands of the Midwest. As a result, the characteristic vegetation of the region represents a mixture of both areas. The region includes approximately 191 million acres of which less than 10% are forested. Precipitation is highly variable across the region. The average annual rainfall ranges from a low of less than 20 inches per year in the North to a high of approximately 55 inches per year along the Gulf Coast of Texas. Many of the areas of the region are marginal for tree growth.

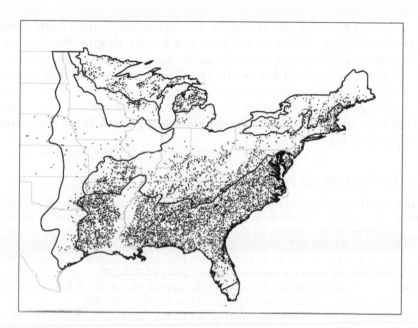

Figure 3.10 Distribution of oak/pine forest types as sampled by the USDA Forest Service Forest Inventory and Analysis Program.

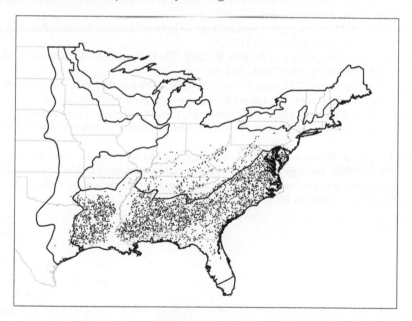

Figure 3.11 Distribution of the loblolly pine/hardwood and Virginia pine/southern red oak forest types as sampled by the USDA Forest Service Forest Inventory and Analysis Program.

Repeated burning and droughty conditions have historically been major disturbances within the region and largely shaped the few forests in the region today. Currently, agriculture is a common land use. Agriculture has replaced fire as the restrictive agent in the region, and forests are relegated to riparian zones and slopes not suitable for agricultural production (Johnson et al. 2002). Post oak and blackjack oak dominate in the southern portion of the region, while bur oak is the dominant oak species in the North.

Further suggested reading

Abrams, M.D. 1992. Fire and the development of oak forests. *Bioscience* 42: 346–353.

Abrams, M.D. 1996. Distribution, historical development and ecophysiological attributes of oak species in the eastern United States. *Annales des Sciences Forestieres* 53: 487–512.

Aizen M.A. and Patterson, W.A. 1990. Acorn size and geographical range in the North American oaks. *Journal of Biogeography* 17: 327–332.

Beilmann, A.P. and Brenner, L.G. 1951. The recent intrusion of forests in the Ozarks. *Annals of the Missouri Botanical Garden* 38(3): 261–282.

Eyre, F.H. (ed.). 1980. *Forest Cover Types of the United States and Canada*. Society of American Foresters, Washington, DC, 148 pp.

Hodges, J.D. 1995. The southern bottomland hardwood region and brown loam bluffs subregion. In: Barrett, J.W., ed. *Regional Silviculture of the United States*. John Wiley, New York, pp. 227–267.

Hodges, J.D., 1997. Development and ecology of bottomland hardwood sites. *Forest Ecology and Management* 90: 117–125.

Johnson, P.S., Shifley, S.R., and Rogers, R. 2002. *The Ecology and Silviculture of Oaks*. CABI Publishing, New York, 544 pp.

Landers, J.L., van Lear, D.H., and Boyer, W.D. 1995. The longleaf pine forests of the southeast: Requiem or renaissance? *Journal of Forestry* 93(11): 39–44.

Little, S. 1979. Fire and plant succession in the New Jersey Pine Barrens. In: Forman, R.T.T. *Pine Barrens: Ecosystem and Landscape*. Academic Press, New York.

Ober, H.K. and Minogue, P.J. 2008. Managing oaks to produce food for wildlife. Wildlife Ecology and Conservation Department, Florida Cooperative Extension Service, Institute of Food and Agricultural Sciences, University of Florida. Publication WEC249.

chapter four

Oak forests' value to wildlife

Todd Fearer

Contents

Acorn production characteristics 42
Use of acorns and oaks by wildlife 42
    Black bears 42
    White-tailed deer 43
    Small mammals 44
    Eastern wild turkeys 44
    Ruffed grouse 45
    Nongame forest birds 45
No suitable substitute for oaks and acorns 46
Oak forest management and its benefits to wildlife 47
Further suggested reading 48

Acorns serve as food for more than 100 species of wildlife, and many of these rely heavily on acorns during the fall and winter. Whereas acorns are the most obvious food provided by oaks, leaves and buds of oaks are an important source of browse for several wildlife species through the year. In addition to food, oaks also provide cover for a wide variety of wildlife species, ranging from bats to bears.

Oak species can be divided into two broad groups: red oaks and white oaks (see Chapter 5). These divisions reflect differences in the flowering and acorn production characteristics between these two groups. Those in the white oak group produce acorns in the fall from flowers produced on the tree the spring of that year. Species in the red oak group produce acorns from flowers that bloomed during the spring of the previous year. Some common species within the red oak group include the northern red oak, black oak, scarlet oak, southern red oak, and pin oak. Species that are common in the white oak group include white oak, chestnut oak, post oak, and bur oak.

Acorns produced by trees from the red oak group contain more fat than those from the white oak group, whereas acorns of several species within the white oak group are more palatable, or less bitter, than those of the red oak group. This is related to higher tannin levels in red oak acorns. Tannins, which belong to a group of chemical compounds called *phenols*, produce a bitter taste in acorns, especially several of the red oaks. They can also interfere with various digestive and physiological functions in animals. In spite of these differences, acorns from both species groups are excellent sources of energy and are easily digested relative to many other foods consumed by wildlife in the Eastern United States.

Acorn production characteristics

Despite tree-to-tree or stand-to-stand variability, patterns of acorn production over broad regions are apparent. Studies have shown that acorn production patterns from year to year are similar in forest stands separated by 300 miles or more. For example, if an oak stand in Southwestern Virginia produces an abundance of acorns (or very few) in a given year, it is likely that an oak stand in Western Maryland also will produce a large (or small) acorn crop that same year. This broad geographic similarity in acorn production is referred to as synchrony and has been documented within both the red oak and white oak species groups. However, because of the differences in the flowering and acorn production characteristics that exist between these two groups, they rarely follow the same synchronous patterns. These synchronous boom-and-bust patterns in annual acorn production affect wildlife species that rely on them as a major fall and winter food source. For instance, how much (or little) animals move during the fall and winter in search of food, and their success in reproducing and raising young the following spring have both been linked to the size of the acorn crop. These influences ripple through the food chain, indirectly affecting many other species of wildlife and playing an important role in shaping the dominant forest and wildlife dynamics that exist in eastern oak forests.

Use of acorns and oaks by wildlife

Black bears

Black bears, though opportunistic feeders, are a good example of a species that rely on acorns as an important component in their diet. The fall is a critical period for black bears as they enter a period of overeating to accumulate fat and energy in preparation for the winter and reproduction. The high-fat content of acorns makes them a prime fall food source. Acorns can be the bulk of a bear's fall diet when readily available, and no other naturally occurring fall food provides similar levels of fat and energy as acorns. As a result, the fall movement patterns of black bears will change to reflect the availability of acorns. During years of poor acorn production, black bears will move two to four times farther in search of pockets of acorn production or other food sources than in years with good acorn production.

Successful reproduction in black bears is very closely tied to the nutritional condition of female bears in the fall. In parts of the black bear's range where acorns are the primary fall food source, studies have shown during years of high acorn production, the majority of sows will give birth to cubs and the survival of those cubs is relatively high. In years of poor acorn production, the proportion of sows giving birth is much lower, as is the survival of their cubs. Where other fall foods are available in abundance, such as grapes, pokeweed, or agricultural crops (corn, wheat, peanuts), they can provide the nutrition necessary for sows to reproduce successively. However, it is worth noting that many of these other fall food sources tend to vary in their annual production or, in the case of crops, do not have the dominant presence in heavily forested areas as oaks. It is also worth noting that squawroot, a parasitic plant that grows from the roots of oaks, is the most important food for black bears during the spring and early summer in the Southern Appalachians.

In addition to providing food, larger oak trees, typically 36 inches or larger in diameter, can provide den sites for bears, especially in the Southeastern portion of their range. Studies in Arkansas, North Carolina, Tennessee, and Virginia have noted the use of cavities in large oaks, primarily northern red oaks, chestnut oaks, and white oaks, as dens for black bears.

White-tailed deer

White-tailed deer consume a tremendous variety of foods. However, despite their diverse diet, acorns are a very important fall food source for them where their range overlaps with oak-dominated forests. Several studies throughout the eastern portion of their range have shown when acorns are available, whitetails readily consume them more than most other available fall foods, with acorns of the white oak group generally favored slightly over those from the red oak group. Acorns provide the fat and energy that deer need for winter survival, and acorns are especially critical in those parts of the white-tail deer's range where spring and summer foods are limited and/or of poor quality. In years when oaks produce a large acorn crop, white-tailed deer tend to weigh more as they enter the winter months. The number of yearling that produce young the following spring also increases when compared to years with little or no acorn crop. For bucks in the Southern Appalachians, antler length, beam diameter, and the average number of antler points all tend to be greater in the year following a large acorn crop. Like black bears, white-tail deer will seek out acorns and shift their fall movements to frequent oak-dominated stands when good acorn crops are available.

Acorns are not the only food that oak trees provide for white-tailed deer. White-tailed deer also consume the leaves and buds of oak seedlings and saplings when other more preferred foods are not available. However, in areas with high deer densities, intense browsing on oak seedlings and saplings can greatly reduce and even eliminate the regrowth of oaks in a forest stand (see Chapter 17).

Small mammals

Acorns are an important source of food for several species of small mammals. While the most recognized of these are perhaps gray and fox squirrels, a considerable portion of the diets of red squirrels, eastern chipmunks, southern flying squirrels, deer mice, and white-footed mice also includes acorns. Many of these species store, or cache, acorns to make them available during the winter and early spring when few other sources of food are available. Gray and fox squirrels typically cache acorns individually, mainly by burying them underground. Many of these acorns germinate and represent an important source of oak seedlings. Other small mammals, such as red squirrels, cache acorns in larger collections located in underground nests or tree cavities. While some studies suggest hickory nuts and black walnuts tend to be preferred over acorns by some small mammals, especially squirrels, acorns still play a critical role in all of their diets. Indeed, squirrel, chipmunk, white-footed mice, and deer mice populations all increase dramatically in the year following a large acorn crop in response to the extra food available to help them through the winter and improve their reproduction.

In addition to food, oaks provide a variety of cover sources for different species of small mammals, especially during the winter. Though most upland oak species are deciduous, many retain some of their dead leaves through the winter. These leaves can serve as winter roost sites for bats, whereas squirrels and other small mammals use the leaves to insulate their nests during the cold winter months.

Eastern wild turkeys

Acorns are an important source of food for wild turkeys throughout their range and are chosen over most other available foods, even when alternate foods are also abundant. Wild turkeys show no preference for acorns from the white oak group over the red oak group. Instead, the size and shape of the acorn may play a role in which species are most preferred. Smaller acorns more rounded in shape and thus easier to swallow, such as those from scarlet oaks, pin oaks, and white oaks, tend to be preferred over larger acorns with a more irregular shape, such as those from chestnut oaks and swamp white oaks. In spite of these preferences, as different species of oak acorns mature and become available, all will be eaten by wild turkeys. Turkey movement patterns can be affected by the availability of acorns in the fall. It is common to see turkeys shift from a summer range of fields and open areas to woods containing oaks in the fall. These shifts can be very distinct and abrupt, coinciding with when acorns become available on the forest floor. In years when the acorn crop is poor, these shifts between the summer and fall/winter ranges tend to be much longer as turkeys seek out alternate or less available foods, including row crops. Hunters in the East may have a better chance of killing a turkey during the fall season when acorn crops are poor. During these years, turkeys increase their activity around alternate and more concentrated food sources, such as fields, orchards, or row crops, making them more visible to hunters. In years of severe winter weather, the presence of abundant acorns and other winter foods may improve the fall and winter survival of turkeys, especially in the northern and northeastern portions of their range.

Ruffed grouse

The majority of the ruffed grouse's range overlaps with aspen, and aspen buds and leaves serve as a primary food source. However, part of the grouse's range includes the oak forests of the Central and Southern Appalachians where aspen is typically absent. In this part of their range, ruffed grouse rely heavily on acorns as a fall and winter food source. In years of poor acorn production in the Central and Southern Appalachians, ruffed grouse increase their home range size up to two-and-a-half times as they search for alternate food sources. In years when acorn crops are large enough to still be available for food in March and early April, female ruffed grouse have greater fat reserves than females without acorns available to eat, leaving them in better condition for the approaching nesting season. Acorn availability is directly related to the reproductive output of ruffed grouse in the Central and Southern Appalachians.

Nongame forest birds

In the eastern portion of the oak's range, blue jays are the primary nongame bird species that regularly eat and store acorns. Like squirrels, blue jays will cache individual acorns in a variety of locations, such as deep grooves in the bark of a tree or similar nooks and crannies that provide a suitable spot for hiding an acorn. Like turkeys, some studies suggest blue jays prefer smaller acorns, such as those from pin oaks and white oaks, but unlike wild turkeys, blue jays do not eat the larger acorns produced by northern red oaks or chestnut oaks. The main reason for this is likely related to the difficulty blue jays have in swallowing the larger acorns.

Acorn production can indirectly affect several species of forest songbirds. For example, several small mammal species that eat acorns, especially white-footed mice and eastern chipmunks, also prey on eggs and nestling songbirds during the spring and early summer. As mentioned earlier, populations of chipmunks, mice, and other rodents that consume acorns increase in the year following a large acorn crop. Consequently, with more rodents present in the forest, their predation of forest songbird eggs and nestlings increases. Short-term but notable reductions in the populations of some forest songbird species, such as wood thrushes, veeries, and hooded warblers, are evident in the years following large acorn crops. On the other hand, the increased populations of squirrels, chipmunks, and other acorn-consuming rodents following a large acorn crop can benefit hawks and owls that feed on these rodents. The abundance of sharp-shinned and Cooper's hawks has been shown to increase in 2 years after a large acorn crop. This increase comes after receiving an extra boost from the abundant rodents available to eat the previous summer, which also was a function of that bumper acorn crop. Other populations of raptors that prey on these rodents, such as barred owls and broad-winged hawks, follow similar patterns.

No suitable substitute for oaks and acorns

Tree seed crops, or mast, such as acorns, hickory nuts, and cherries, are the most valuable and energy-rich plant food available for wildlife in eastern forests during the fall and winter. At the time of European settlement, the most abundant and widespread mast-producing trees were oaks, beech, hickories, and chestnuts. The American chestnut apparently was the most common nut-producing tree, and the American beech the most widely distributed in the eastern forest. Today, the American chestnut has been virtually eliminated and the abundance of the American beech has decreased dramatically, primarily as a result of introduced diseases and nonnative insects. Consequently, the importance of oaks in eastern forests has increased substantially since the past century, especially in the role they play as food and cover sources for wildlife. Hickories remain abundant, but the hard, woody shell renders hickory nuts less useful to species that cannot open them, as opposed to squirrels, chipmunks, and other rodents. While important to several species of wildlife as food, soft mast species that produce fleshy fruits, such as cherries and serviceberries, do not contain the fat levels critical in building fat reserves for many wildlife species. Therefore, no other tree species common to the eastern forests of the United States fills the multifunctional role of oaks for wildlife. This fact underscores both the necessity to manage forest stands for oaks and methods to ensure their long-term persistence in the forest landscape.

Oak forest management and its benefits to wildlife

Proper and active management of oak forests is necessary to ensure their persistence in the forest landscape (see Chapters 13 through 16). Many oak forests in the Eastern United States are facing a variety of challenges that threaten their long-term sustainability, such as reduced health resulting from insects and diseases, stress from overabundant white-tailed deer populations (see Chapters 6 and 17), and permanent replacement by other forest types as existing oak stands continue to mature and change. Effective management of oak forests, which can include cutting trees, using fire, and other standard silvicultural techniques (see Chapters 7 through 11), is necessary to successfully address these challenges. However, a common misconception is that many of these practices are damaging to the forest and especially the wildlife that live in them. To the contrary, many wildlife species benefit greatly from oak forest management. The species that receive the greatest benefit will depend on many factors, especially the objectives of the management (see Chapter 13) and the current conditions of the forest (see Chapter 12).

Several species of wildlife use a variety of forest age classes to meet their different food and cover needs throughout the year. Other species prefer specific age classes, such as young forests, because they have specific habitat preferences that are only found in that age forest. For example, many bird species prefer young forest stands between approximately 5 and 20 years old. These young stands are easily created by a timber harvest and typically contain a high density of seedlings and saplings, making them very thick and brushy and difficult to walk through. Ruffed grouse and the American woodcock use these young stands as their primary source of cover, and several forest songbirds, including prairie warblers, golden-winged warblers, and eastern towhees, prefer the thick cover provided by these stands for nesting during the spring. Other forest songbirds that nest in mature oak stands, such as cerulean warblers and wood thrushes, will use these younger forest stands to forage with their young later in the summer and prepare for the upcoming migration. Younger forests also provide important sources of browse and cover for deer and rabbits.

Mature stands, approximately 60 years or more old, also are critical for several species of wildlife. As mentioned, many forest songbirds nest in mature stands, wild turkeys prefer the more open understory of these stands, and black bears need mature stands for both food and large cavity trees for denning. The trees in these mature stands produce the majority of acorns. Because wildlife species use a range of forest age classes, the most beneficial forest management plans are those that create a diversity of forest age classes in the landscape.

A common decision faced by many landowners in managing their oak forest is the apparent dilemma of harvesting trees to provide revenue, create young forest cover for wildlife, or regenerate their aging stands versus maintaining mature oaks for acorn production. This decision is partly dependent on the landowner's long-term objectives, the acreage of land available for management, and should be discussed with both an experienced forest manager as well as a wildlife biologist. It should also involve knowledge of the ecology of oak forests and looking at the forests in the surrounding area.

Eastern oak forests, especially upland forests, are considered disturbance dependent. That is, a mature oak forest requires some type of disturbance for oak seedlings and saplings to successively become established in the forest understory. If left unmanaged, it is likely that most oak stands will slowly transition to other tree species that are more shade tolerant, such as maples and the American beech (see Chapter 6). These disturbances can be natural, such as a large storm or fire. However, it is more common for them to be artificially created by harvesting trees using a variety of techniques (see Chapter 7). While this reduces the number of oaks and acorns in a stand and affects wildlife species that use them for cover and food, it is a necessary step in the long-term maintenance of the oak forest.

A critical step in deciding to harvest trees versus keeping the mature trees for acorn production is to not consider a single stand or property in a vacuum. Instead, look at the surrounding stands and consider their condition as well as any active management occurring in them. If managing your forest to maximize benefits to wildlife is one of your primary objectives, remember wildlife does not recognize property boundaries. If a neighboring landowner is harvesting a lot of timber, then maintaining mature stands on your property may provide the best benefit for wildlife. Conversely, if your property is surrounded by mature oak forests with no plans for harvest, they will provide the acorn production and harvesting the timber on your property will create young forest structure highly beneficial for wildlife.

Further suggested reading

Dessecker, D., G. Norman, and S. Williamson (eds.). 2006. Ruffed grouse conservation plan. Association of Fish and Wildlife Agencies, Resident Game Bird Working Group (available at: www.ruffedgrousesociety.org/UserFiles/File/RG_ConservationPlan.pdf, accessed September 18, 2015.)

Dickson, J. 2004. Wildlife and upland oak forests, in Spetich, MA (eds.) *Upland Oak Ecology Symposium: History, Current Conditions, and Sustainability*, General Technical Report SRS-73, pp. 106–115. U.S. Department of Agriculture, Forest Service, Southern Research Station, Asheville, NC, 311 pp. (available at: www.treesearch.fs.fed.us/pubs/6506, accessed September 18, 2015.)

McShea, W. and W. Healy (eds.). 2002. *Oak Forest Ecosystems: Ecology and Management for Wildlife*. The John Hopkins University Press, Baltimore, MD, 431 pp.

Stauffer, D.S., J.W. Edwards, W.M. Giuliano, and G.W. Norman (eds.). 2011. *Ecology and Management of Appalachian Ruffed Grouse*. Hancock House Publishers, Blaine, WA, 176 pp.

chapter five

Silvics of oaks
Their biology and ecology

Thomas R. Fox and Jerre Creighton

Contents

Silvics and forest management 53
Seed production 55
Sprouting ability.. 56
Shade tolerance.. 57
Tolerance of poor soil drainage and flooding 58
Fire tolerance.. 58
Growth rate 59
Longevity 59
Insects and diseases 60
Summary and conclusion 60
Further suggested reading.. 61

The hardwood forests of the Eastern United States are among the most diverse forests in the world. Oaks are an important component in many of these forests, but they seldom grow in pure stands anywhere in the region. Most forests in the Eastern United States contain dozens of species in addition to oaks (see Chapter 3). For example, in the hardwood forests of the Southern Appalachians, it is possible to group forests into four general types, based on the species that commonly occur and differences in the growth rates of the forest (Table 5.1). Using this approach, the mix of trees in a white oak/black oak forest is different than the mix of trees found in a sugar maple/northern red oak forest. Of course, some types of trees, such as white oak or red maple, might be common in both forest types. In addition, we would expect the trees in the white oak/black oak forest would grow slower than the trees found in a sugar maple/northern hardwood forest.

Different types of forests occur on different parts of the landscape, and their growth rates differ because of temperature, rainfall, and soils. As a result, the same species of trees commonly occur in different areas on the landscape that share similarities in these characteristics. In the

Table 5.1 Major hardwood forest types described by Smith (1994) in the Southern Appalachian Mountains, including the site index range, typical growth rate, and species commonly found in each type of forest

Scarlet oak–chestnut oak forest	White oak–black oak forest	Sugar maple–northern red oak forest	Yellow poplar–mixed hardwood forest
Site index range			
<55 feet base age 50	56–70 feet base age 50	70–85 feet base age 50	>85 feet base age 50
Growth rate in mature fully stocked stands			
150 board feet/ acre/year	235 board feet/ acre/year	320 board feet/ acre/year	640 board feet/ acre/year
Common trees			
Scarlet oak	White oak	Northern red oak	Yellow poplar
Chestnut oak	Black oak	Sugar maple	Northern red oak
Virginia pine	Scarlet oak	Yellow poplar	Sugar maple
Pitch pine	Chestnut oak	Black cherry	White ash
Table mountain	Pignut hickory	White ash	Basswood
Pine	Mockernut hickory	Basswood	American beech
Sassafras	Shagbark hickory	American beech	White oak
Bear oak	Red maple	White oak	Cucumbertree
Blackgum	Blackgum	Cucumbertree	Black cherry
	American beech	Red maple	Red maple
	Sweet birch	Frasier magnolia	Frasier magnolia
		Sweet birch	Sweet birch
		Yellow buckeye	Yellow buckeye
		American elm	American elm
		Black walnut	Black walnut
		Eastern white pine	Eastern white pine
		Eastern hemlock	Eastern hemlock

Southern Appalachian Mountains, scarlet oak occurs most often on dry, southwest-facing ridges, white oak on side slopes, and northern red oak in moist, cool coves. Similar patterns of tree distribution occur in bottom-land hardwood forests of the South. Overcup oak occurs in places that are subject to frequent and prolonged flooding, whereas other oaks, such as Nuttall and cherrybark, occur on ridges, levees, and other elevated locations in the floodplain that seldom are under water.

Silvics and forest management

Silviculture is the part of forest management that deals with growing, tending, and harvesting trees in existing forests and establishing new forests following harvest or other disturbances, such as wind damage or wildfire. Sound silvicultural practices are based on the characteristics of the individual trees in a forest, including how they regenerate, grow, and compete with one another, which is particularly important in hardwood forests where planting is rarely done and foresters usually rely on natural regeneration from seed or sprouts following harvest.

Silvics is the study of trees and how they grow. Knowledge of silvics can help explain why different trees occur in different types of forests, why their growth rates vary, why some trees die and other trees live, and what conditions are needed to successfully establish different types of trees after harvest.

Some of the more important silvical characteristics of trees include the following:

- Extent of natural range
- Common associated trees
- Seed production and distribution
- Shade tolerance
- Tolerance of poor soil drainage or flooding
- Fire tolerance
- Growth rate
- Longevity
- Susceptibility to insects, diseases, and other damaging agents such as wind or ice

Table 5.2 lists some of the more common trees found in the Eastern United States and summarizes their important silvical characteristics. For a complete description of the silvics of trees in the United States, see Burns and Honkela (1990) or visit http://www.na.fs.fed.us/spfo/pubs/silvics_ manual/table_of_contents.htm. These references provide a wealth of information on individual species, including their natural range and common associates. We provide information on key silvical issues important to oaks and their management in oak forests in the following sections.

Table 5.2 Silvics of selected tree species found in eastern hardwood forests

Common name[a]	Shade tolerance[b]	Growth rate[c]	Life span[d]	Tolerates poor soil aeration
Common Oaks				
Black oak (R)	I	I	I	No
Cherrybark oak (R)	NT	F	I	Yes
Chestnut oak (W)	I	S	L	No
Laurel oak (R)	T	F	I	Yes
Live oak (R)	I	F	L	Yes
Northern red oak (R)	I	F	L	No
Nuttall oak (R)	I	F	I	Yes
Overcup oak (W)	I	S	VL	Yes
Pin oak (R)	NT	F	I	Yes
Post oak (W)	NT	S	L	No
Scarlet oak (R)	NT	F	I	No
Southern red oak (R)	I	I	I	No
Swamp white oak (W)	I	F	L	Yes
Turkey oak (R)	NT	F	I	No
Water oak (R)	NT	S	I	Yes
White oak (W)	I	S	VL	No
Willow oak (R)	I	F	L	Yes
Other common hardwood species				
American basswood	T	F	I	No
American beech	T	S	VL	No
American holly	T	S	I	Yes
American sycamore	I	F	VL	Yes
Bitternut hickory	I	S	VL	Yes
Black birch	I	I	I	No
Black cherry	NT	F	I	No
Black locust	NT	F	S	No
Black walnut	NT	F	I	No
Blackgum	T	F	I	Yes
Butternut	NT	F	I	No
Cucumbertree	I	F	I	No
Eastern cottonwood	NT	F	I	Yes
Flowering dogwood	T	F	S	No
Frasier magnolia	I	I	I	Yes
Green ash	I	I	I	Yes
Mockernut hickory	NT	S	VL	No

(Continued)

Table 5.2 (Continued) Silvics of selected tree species found in eastern hardwood forests

Common name[a]	Shade tolerance[b]	Growth rate[c]	Life span[d]	Tolerates poor soil aeration
Persimmon	T	S	I	No
Pignut hickory	NT	S	L	No
Red maple	T	F	I	Yes
Sassafras	NT	I	I	No
Shagbark hickory	I	S	L	No
Striped maple	T	S	S	Yes
Sugar maple	T	I	VL	No
Swamp tupelo	NT	I	L	Yes
Sweetgum	NT	F	I	Yes
Trembling aspen	NT	F	S	No
White ash	I	I	I	No
Yellow birch	I	I	L	No
Yellow buckeye	T	I	L	No
Yellow poplar	NT	F	L	Yes

[a] R, red oak group; W, white oak group.
[b] T, tolerant; I, intermediate; NT, intolerant.
[c] S, slow; I, intermediate; F, fast.
[d] S, short (<100 years); I, intermediate (100–200 years); L, long (200–300 years); VL, very long (>300 years).

Seed production

Acorns produced by oaks are the source of new oak seedlings in the forest and an important food source for wildlife. Oaks are divided into two broad groups based on the time it takes for acorns to ripen and when the acorns germinate. In the white oak group, which includes white, chestnut, swamp white, overcup, and swamp chestnut oaks, the acorns ripen in the fall of the same growing season in which the flowers were pollinated. Acorns from this group usually germinate in the fall, soon after they drop from the tree. In the red oak group, which includes northern red, scarlet, black, Nuttall, and water oaks, acorns do not ripen until the fall of the second year after the flowers are pollinated in the spring. Acorns in these species usually overwinter on the forest floor after they drop in the fall and germinate the following spring.

Unlike many species with small seeds that germinate best on mineral soil, acorns germinate best when they fall on the forest floor and are buried by leaves. Acorns typically fall just before the leaves of the oaks drop in autumn, which helps to bury them in the duff layer on the forest floor. Acorns of the red oak group require a few months of cold and wet conditions, called stratification, before they germinate. That is why these acorns drop in the fall but do not germinate until the following spring. In contrast, acorns of the white oak group do not require this period of cold, wet conditions. They germinate in the autumn, soon after they fall.

Acorn production is quite variable in the oaks. Several species, such as Nuttall oak and willow oak, tend to produce good acorn crops every year. Most species only produce large crops of acorns every few years. For example, the northern red oak and white oak produce good acorn crops every 2–5 years, with bumper crops perhaps every 4–10 years. Acorn production in the water oak seems to alternate between good years and poor years. There is also a considerable variation, even in years with good acorn crops. Individual trees may produce huge numbers of acorns, while a nearby tree that looks similar produces almost none (see Chapter 4).

The size of the acorn crop is important because it affects the amount of food available for wildlife and the potential for new oak seedlings to develop. More than 500 acorns may need to fall to the forest floor in order to produce one new northern red oak seedling the following spring. Deer, turkeys, squirrels, rodents, birds, and insects consume large numbers of acorns. In most years, perhaps 80% of the acorns produced by a northern red oak are consumed. In years with poor acorn crops, practically all acorns are eaten. However, many of the rodents, birds, and other animals that consume acorns also help to distribute them throughout the forest (see Chapter 4).

Sprouting ability

Hardwoods in general and oaks in particular often regenerate following harvest or other natural disturbance by sprouting from the stump. Stump sprouts generally grow much faster than newly germinated seedlings, particularly in the oaks where height growth of newly germinated seedlings is very slow. Oak seedlings put more of their effort into growing roots than stems or branches. So the seedlings of other species often grow much faster in height than oak seedlings. Because most oaks do not tolerate shade, seedlings of the faster growing trees often overtop the oaks. On the other hand, the large, well-established root system of a stump sprout provides a large food reserve and is able to acquire large amounts of water and nutrients that enable the sprouts to grow rapidly. If it were not for sprouts from cut stumps, oaks would be less common in many forests.

Not all trees sprout from the stump when they are cut. There are many factors that affect the likelihood of sprouting after a tree is cut, including species, size, and the season when the tree is cut. Although almost all hardwoods will sprout, some sprout more commonly and vigorously than others. For example, yellow poplar sprouts extremely well, and multiple sprouts from a single stump frequently produce multiple sawtimber-size stems, all of which originated from the same stump. Among the oaks commonly found in the Appalachians, scarlet and northern red tend to sprout much more vigorously than white and black oak. Regardless of species, sprouting declines as trees get older and larger. For example, in 40-year-old white oaks, 63% of stumps less than 5 inches in diameter will sprout, but only 9% of stumps 12–16 inches in diameter will sprout. By the time white oaks reach 100 years of age, only 8% of stumps less than 5 inches and 3% of stumps 12–16 inches in diameter will sprout. Trees cut in the spring, just after the leaves come out, sprout less than trees cut in the winter because carbohydrate reserves in the roots are depleted because they have been used to produce new leaves.

Shade tolerance

Shade tolerance describes a tree's ability to survive and grow in the shade of other trees. Trees that are shade intolerant must have nearly full sunlight to survive and grow. Seeds of shade-intolerant trees often do not germinate in the understory, and those that do die very quickly. So seedlings of these species are seldom found in the understory of closed-canopy forests where little direct sunlight reaches the forest floor. Examples of common eastern hardwood trees that are intolerant of shade are scarlet oak, yellow poplar, black cherry, mockernut hickory, and sweetgum. In contrast, shade-tolerant trees are able to germinate, survive, and grow reasonably well in the shade of other trees. Seedlings of these trees are common in the understory of closed-canopy forests and will often form a dense understory of young trees that persist for many years. Examples of shade-tolerant trees include the sugar maple, basswood, flowering dogwood, American beech, blackgum, and persimmon. Except for laurel oak in the Coastal Plain, none of the common oaks in the eastern hardwood forests are considered shade tolerant.

In between the shade-intolerant and the shade-tolerant trees are a large group of trees, including most oaks, which are intermediate in shade tolerance. Seedlings of trees that are intermediate in shade tolerance can survive for some period of time in the understory, but eventually they will need more direct sunlight to survive. If their seeds germinate in the understory of an existing forest where there is substantial shade, or if other trees grow faster and gradually overtop them, shade-intermediate seedlings in the understory will eventually die. This explains why, after a good acorn crop, large numbers of oak seedlings can be found in the understory, but after only a few years, they may be quite sparse. Examples of shade-intermediate trees include white oak, chestnut oak, northern red oak, southern red oak, water oak, green ash, yellow birch, and sycamore.

Tolerance of poor soil drainage and flooding

Hardwoods trees, including oaks, occur on a wide range of soil types, ranging from wet, poorly drained soils near rivers and streams that are frequently flooded, to dry, well-drained soils in uplands and mountainous areas. However, the species on these different types of soils are usually different because of their varying adaptations to soil drainage.

When soils are saturated with water for long periods of time, especially when trees are actively growing, the roots of many species do not grow well. Wet soil is low in oxygen, which tree roots need to survive. Some tree species are able to tolerate wet conditions better than others, enabling them to survive and grow, while the others die. Water tupelo is an example of a hardwood species that can grow in very wet conditions, often on sites that are almost permanently flooded. Among the oaks, overcup oak is the most tolerant of flooding and poor soil drainage. Other oaks, such as Nuttall, willow, water, swamp chestnut, and swamp white oak, tolerate wet soils and are found frequently in bottomlands.

Other oaks that occur in bottomlands but need better drainage, such as cherrybark oak, grow on ridges, levees, and other elevated bottomland sites. In contrast, white, black, northern red, and southern red oaks are most commonly found on well-drained sites where soil moisture is available, but the soils are seldom saturated with water. Scarlet, chestnut, and turkey oaks survive and grow on very dry and droughty soils, including shallow, rocky soils in the mountains and deep sandy soils in the Sandhills of the upper Coastal Plain.

Fire tolerance

Several species of oak depend on fire for successful regeneration and generally tolerate fire better than other hardwood species that typically grow with them, such as red maple or yellow poplar. When there are frequent low-intensity fires in hardwood forests, the proportion of oaks generally increases over time, and the proportion of other species, such as red maple and yellow poplar, often declines.

The ability of trees to tolerate fire depends on the age of the tree, thickness of its bark, and the intensity of the fire (see Chapter 9). Young seedlings tend to have thin bark and are often killed by even relatively low-intensity fires. Sometimes fire will only "top-kill" a seedling or small sapling, leaving the rootstock unharmed. Top-killed stems readily re-sprout. One reason oak seedlings are more fire tolerant than many of their competitors is that they allocate a larger proportion of their growth to the roots. Therefore, oak seedlings generally have a larger root system and, in turn, greater food reserves, which enable them to survive and sprout more aggressively following a fire. This adaptation enables them to survive repeated low-intensity fire that only top-kills the seedlings. More intense fire can kill a seedling's rootstock, which precludes any re-sprouting.

Because of the germination strategy mentioned earlier (buried in the forest duff), oaks have root collars (the part of the tree most vulnerable to fire) situated deeper in the soil than other hardwood species. The extra depth allows oak seedlings to survive fires that kill seedlings of many other hardwood species.

The proportion of oak seedlings and saplings often increases over time in hardwood forests where frequent low-intensity fires occur. As trees get older, they develop thicker bark, which acts as an insulation layer that protects them from fire. Other species with a thinner bark, such as the American beech or maples, may be injured or killed by the same fire. Even if the fire does not directly kill the tree, the wounds created in trees with a thinner bark may allow insects and diseases to enter and can eventually kill the tree.

Of course, a fire that is hot enough may kill all the trees in a forest. Fortunately, high-intensity fire is rare in most hardwood forests in the Eastern United States.

Growth rate

The growth rate of trees is determined by rainfall, temperature, the amount of water and nutrients in the soil, competition from other trees and shrubs, and the impact of damaging agents, such as insects and diseases. As with any plant, most trees grow faster when on deep, fertile soils, receive adequate rainfall, and are free from insects and diseases. However, there are substantial differences among species of trees in their inherent growth potential. Even when growing on the same soil and in the same climate, some trees will grow faster than others. Among the oaks, northern red, scarlet, cherrybark, Nuttall, pin, and willow are considered fast-growing species, whereas white, overcup, and chestnut oaks grow more slowly.

Longevity

Following a major disturbance, such as a fire, windstorm, or harvest, tens of thousands of small seedlings may germinate on each acre of the forest floor. It is one of the wonderful things that make our hardwood forests so resilient. These seedlings compete with one another for water, nutrients, and light. As these seedlings grow larger, many will die at a young age as a result of competition from other trees, part of the natural process known as *succession*. So many seedlings and saplings are killed by competition that by the time a forest matures, there will be no more than a few hundred trees remaining on each acre. Surviving trees may live for decades or centuries.

Individual mature trees may die because of insects, diseases, lightning, fires, windstorms, or other natural processes, but eventually all trees die. Oaks are among the longest-living species in the eastern hardwood forest. White and chestnut oaks may live for 400 years or more. Most oaks have shorter life spans, up to 200 or 300 years for species such as northern red, black, and cherrybark oaks. Some species are naturally much shorter lived; scarlet and turkey oaks seldom live past 100 years.

Insects and diseases

Trees are under almost constant attack from various insects and diseases. Some of these pests have only minor impacts on the trees, whereas others may kill trees over thousands of acres. The typical longevity of many trees is often determined by their susceptibility to certain insects and diseases.

The hardwood trees of the Eastern United States have coexisted for thousands of years with insects and diseases native to the region, so they tend to be somewhat resistant to them. As a result, native pests tend to have a relatively small impact on oaks. Growth rates may be reduced and some weaker trees killed, but the overall effect is small.

Nonnative insects and diseases—introduced from other continents— have had a much more dramatic effect because the native trees have little or no resistance. Chestnut blight is a classic example of an introduced disease, one that dramatically changed the eastern hardwood forest and has virtually eliminated one of the most important hardwood trees in the entire United States (see Chapter 2).

Among the several introduced insects severely impacting the eastern hardwood forest is the gypsy moth, which came to Northeastern United States around 1869 and has spread throughout much of the Eastern United States, impacting hundreds of thousands of acres. Oaks are among the preferred foods of the gypsy moth and are very susceptible to attack.

Summary and conclusion

Hardwood forests cover extensive areas in the Eastern United States. Oaks are a significant component of many of these forests, though a wide variety of other species are usually present as well. Trees vary in many important properties, such as shade tolerance, longevity, and growth rate. In order to make a sound silvicultural prescription, whether the objective is to grow timber, improve the stand for wildlife increase the production of nontimber forest products, or improve recreation opportunities, a thorough knowledge of silvics is required. Armed with this knowledge, landowners are better able to develop and implement sound silvicultural plans that will enable them to successfully manage their forests and accomplish their objectives.

Further suggested reading

Burns, R.M., and B.H. Honkala (tech. coords). 1990. *Silvics of North America, Vol. 1: Conifers; Vol. 2: Hardwoods, Agriculture Handbook 654*. U.S. Department of Agriculture, Forest Service, Washington, DC, Vol. 2, 877 pp. Available online: http://www.na.fs.fed.us/spfo/pubs/silvics_manual/table_of_contents.htm, accessed August 27, 2015.

Hodges, J.D. 1997. Development and ecology of bottomland hardwood sites. *Forest Ecology and Management*. 90(2–3):117–125.

Johnson, P.S. 1993. *Perspectives on the Ecology and Silviculture of Oak-Dominated Forests in the Central and Eastern States*, General Technical Report NC-153. U.S. Department of Agriculture, Forest Service, St Paul, MN, 28 pp.

Smith, D.W. 1994. The southern Appalachian hardwood region (Chapter 5). In: *Regional Silviculture of the United States*, 3rd edn., edited by J.W. Barrett, John Wiley & Sons, Inc. New York.

chapter six

Oak regeneration challenges

Jeffrey W. Stringer

Contents

Principles of oak regeneration.. 64
Historic and current oak regeneration 66
Site quality and oak regeneration.. 68
Assessing regeneration potential... 68
Advance regeneration 69
Stump sprouts.. 69
Management challenges.. 70
Further suggested reading... 71

Lack of adequate oak regeneration in the face of increasing competition from other tree species has been recognized as a significant problem for many oak-dominated forests for several decades and has been implicated in the slow loss of oaks from our forests. This recognition has led to a considerable amount of research devoted to understanding where and why this is occurring. Although research has yielded results to better understand how to deal with oak regeneration problems, it is still a challenge in many forest types to regenerate oak to desired levels.

Techniques have been developed to successfully regenerate oak. However, many of these involve several silvicultural treatments conducted over a number of years. Implementing several treatments over a long period of time is challenging compared to conducting a one-time operation. In contrast to oaks, yellow poplar stands can easily be regenerated through a single harvest where the majority of trees are removed. This harvest disturbs the layer of leaves on the ground, exposes mineral soil, and at the same time removes shade from the site. These conditions, coupled with the seed that has built up in the duff layer, generally result in plentiful yellow poplar seedlings. These seedlings, which grow quickly, can easily result in stands of nearly 100% yellow poplar. This is similar to regenerating other species that grow rapidly from seed, such as southern pines. This means you do not have to think about regenerating these species until there is a significant overstory harvest to develop a regenerating age class.

However, this is not the case with oaks. A single treatment, such as a clearcut or traditional shelterwood (see Chapter 7), in oak-dominated stands on relatively high-quality sites generally results in oak regeneration failures because the development of oak regeneration is a process that occurs over 5–20 years. This time frame can be a challenge to many forest owners.

To effectively develop a plan for regenerating oaks, it is important to understand the process of oak regeneration and how to implement treatments to encourage its development.

Principles of oak regeneration

There are several important principles of oak biology that must be understood to effectively manage and regenerate oak. The most important consideration for oak regeneration is for adequate oak reproduction to already be established prior to a regeneration harvest or treatment. Proper reproduction can come in two forms: (1) stump sprouts (see Chapter 5) and/or (2) seedlings or saplings that possess enough vigor to grow rapidly upon release.

The latter is termed advanced oak regeneration. That is, young trees present before or in advance of a regeneration harvest or treatment. Without trees that can be cut and expected to sprout or the presence of advanced oak regeneration, oak will not successfully regenerate in a newly developing stand. An acorn present at the time of harvest has no chance of germinating and growing quickly enough to keep up with fast-growing competitors. A good example is competition from yellow poplar, which can generate a fast-growing tree from seed that can easily grow 10 feet tall within 3–4 years after harvest. It can do this because it puts most of its resources (water, nutrients, and food it generates from photosynthesis) into increasing height. In contrast, oaks initially spend time developing a large root system instead of height growth. The result is that yellow poplar can tower above oak seedlings and eventually shade them out. The relative slower height growth of oak seedlings compared to competing species is a problem and can prevent both the natural regeneration of oak and efforts to establish oak from planted seedlings and acorns. However, if you cut oak trees that are capable of sprouting (producing stump sprouts) or have large advanced oak regeneration, these can develop stems that can grow rapidly in height and compete successfully with competitors that arise from seed (e.g., yellow poplar). The more numerous the stump sprouts and/or the larger in size the advanced oak regeneration, the more likely oak regeneration will be successful.

It is assumed that competition from species that commonly are found with oaks will occur. It is unusual for this not to happen. It is also difficult and expensive to eradicate competing species from a stand and keep them from seeding into or sprouting back in a regenerating stand. Therefore, efforts to successfully regenerate oak have focused on developing vigorous advanced oak regeneration or stump sprouts coupled with some competition control.

Figure 6.1 shows the likelihood the different types of oak reproduction will produce a tree that will grow into a canopy tree in a regenerating age class. This information allows you to evaluate your forests to determine if oaks have the potential to regenerate and if management is needed to regenerate oaks. Unfortunately, many oak forests, especially those on medium- to high-quality sites, have little advanced oak regeneration and few pole-sized trees that can develop sprouts, which gives little hope of maintaining oak after a typical regeneration harvest. In some cases, there may be patches of small seedlings, maybe a foot tall. However, these seedlings are not large enough to quickly respond to the release of a regeneration harvest with rapid height growth and will not outpace their competitors.

The reason behind the failure is largely governed by shade tolerance. As discussed in Chapter 5, oak is generally considered intermediate in shade tolerance. That is, oaks can tolerate some shade but not heavy shade. When an acorn germinates in heavy shade, it grows slowly in height and eventually dies if shady conditions persist, which explains why there are so few large (3–4 feet tall and greater) oak seedlings in forests on medium- to high-quality sites—there is too much shade. On more productive sites, it is common to find shade-tolerant species that

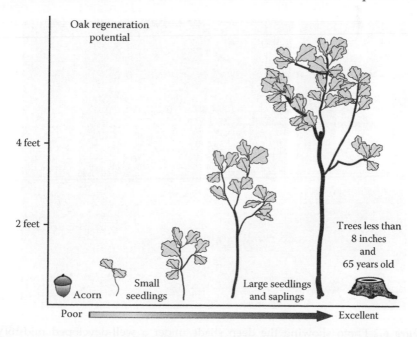

Figure 6.1 The likelihood of oak reproduction to develop into a canopy tree after an even-aged regeneration harvest.

proliferate as mid- and understory trees producing heavy shade that further suppress oak seedlings. It is common for these forests to repeat the process of sporadically producing a bumper crop of acorns that germinate into seedlings, then decrease in number over time in the presence of the heavily shaded understory. Under these conditions, there is little hope of developing the number of advanced regeneration seedlings needed to perpetuate oak.

Historic and current oak regeneration

The scenario described earlier begs the question of how oaks ever became dominant overstory trees. The answer is that conditions present when the overstory trees were only seedlings or saplings were far different than they are now. Clearly, the disturbances discussed in Chapter 2 created more light on the forest floor and reduced the number of competing species, including those that perpetuate themselves as understory and midstory trees. These understory and midstory species (such as the sugar and red maple and American beech) are shade tolerant and can develop very thick understories and midstories, thus shading out the oak regeneration. Figure 6.2 shows the deep shade created by these shade-tolerant midstory trees. In the foreground, red and sugar maples have been cut from the

Figure 6.2 Photo showing the deep shade under a well-developed midstory compared to the high level of diffuse light once the midstory has been removed (foreground).

midstory resulting in light shading where small oak seedlings can now thrive. This is in contrast to the dark conditions found in the forest shown in the background where no midstory trees were removed. These midstory species also have relatively thin bark and are susceptible to damage by fire. Past land disturbances, especially from a long-term fire regime, would result in a reduction in small shade-tolerant trees over time. The reduction in these species improves light conditions for establishing large numbers of oak seedlings and encourages successful regeneration of forests rich in oaks that we see today. Currently, there are fewer of these disturbances and competing species have reestablished themselves in oak forests, resulting in dense understories and midstories. The resulting leaf area decreases light at seedling height to an unacceptable level, preventing oak regeneration. This produces the common situation of oak in the overstory with inadequate advanced oak regeneration and little hope of regeneration without something changing.

To regenerate oak forests, there must be adequate sources of regeneration coupled with methods that create favorable light conditions. To start the process of establishing advanced oak regeneration, you need an abundant acorn crop. Bumper crops are needed so that some of the acorns escape being eaten by wildlife and insects. Sometimes this wait can be a challenge to management. Once seedlings have developed, silvicultural treatments are prescribed to ensure proper light conditions that will enable the seedlings to develop into vigorous advanced regeneration. Figure 6.3 shows examples of oak seedlings of various degrees of vigor and release response.

(a)   (b)   (c)

Figure 6.3 Three advance regeneration examples: (a) small 1-foot-tall, 14-year-old white oak seedling growing on a medium-quality site under a red maple understory, (b) 2½-foot-tall red oak seedling, and (c) vigorous 4- to 5-feet-tall northern red oak advance regeneration capable of responding well to overstory removal.

Site quality and oak regeneration

As previously mentioned, there is a relationship between successful oak regeneration and site quality. On poor-quality sites, oaks can be expected to regenerate without the problems encountered on high-quality sites. This is because of a suite of characteristics that combine to either favor or disfavor oak. Figure 6.4 shows the relationship of a number of important characteristics that affect oak regeneration as related to site quality. As site quality increases, so does the overall leaf area of a stand. This is from the leaf area of the canopy, as well as the understory and midstory, which ultimately decreases light infiltration to the forest floor. Also, as site quality increases, so do the number of competing species and the height growth potential of the competing species. The end result is high-quality sites produce more shade, leading to low numbers of small, advanced oak regeneration and significant numbers of competitors that grow quickly after a regeneration harvest.

Assessing regeneration potential

The challenge of regenerating oak lies in the development of stump sprouts and advanced oak regeneration. It is helpful to have an understanding of what constitutes appropriately sized advanced regeneration, and which trees can produce abundant sprouts when cut. Although the appropriate

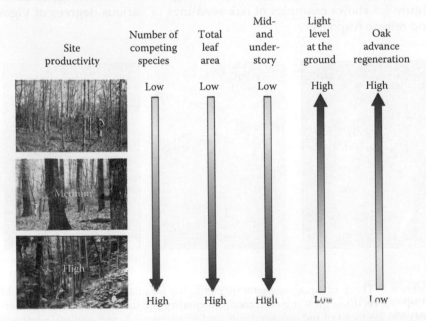

Figure 6.4 Characteristics associated with oak success relative to site quality.

size and number depend on the oak species, site characteristics, and shade competition from other trees, there are some general guidelines to assess regeneration potential.

Advance regeneration

Advance oak regeneration must have enough vigor to develop rapid height growth when released by a regeneration treatment. Generally, height is a good indicator of advance regeneration vigor. Small oak seedlings that have persisted in the understory for a long period of time with little change in their height indicate suppression. They are barely surviving in shaded understory conditions and do not have the resources necessary to respond to release (Figure 6.3a). Taller seedlings indicate there has been enough light present to fulfill growth needs and build larger root systems with a storehouse of resources that translate into growth. Typically, vigorous advance regeneration is indicated by a height of approximately 4 feet and a crown that supports abundant leaves (Figure 6.3c). Seedlings of this size and larger are needed on sites that have high growth potential and significant competition from other species. Shorter seedlings have been known to respond well and produce canopy trees on relatively poor-quality sites with fewer competitors or in the case of fast-growing oaks such as cherrybark on rich bottomland sites with minimal ground and understory vegetation. Typically, upland sites of medium to high productivity, where competing species are present, need 4 foot or taller advance oak regeneration. It is important to have 100–200 or more per acre, to ensure adequate oak in regenerating age classes.

Stump sprouts

Trees greater than 1–2 inches in diameter are beyond the size normally considered oak regeneration. However, trees of this diameter and those of sapling and pole size (2–10 inches) may sprout prolifically upon removal of their top, either through intentional cutting, logging damage, or fire. Removal of the top allows suppressed buds that occur at the groundline to produce stump sprouts, also called *coppice sprouts*. The number and height growth of the sprouts from an individual tree that has the top severed are related to the number of suppressed buds at the base of the stem (root collar) and the size and carbohydrate supply of the root system. The abundance of suppressed buds and favorable root characteristics are based on species and the vigor of the sapling or pole-sized tree. Age is a factor that comes into play when assessing vigor. The condition of a tree's crown can also be used to indicate its vigor. Generally, red oaks, which are less shade tolerant than white oaks, tend to lose vigor and suppressed buds at 30 years of age when suppressed in the understory or midstory. These trees often will have crowns that have lost major branches or are very small and flat topped. White oaks can maintain vigor for a longer period of time, typically 60 years or more. However, when white oak sapling or pole-sized trees have been living under an intact canopy for 80 years, they sprout little when cut. Again, these stems typically have similar crown characteristics as the suppressed red oak. If you have younger trees, they generally will sprout. Having at least 20 of these stems per acre will provide oak in regenerating stands. Stump sprouts have a high probability of creating stems that will grow fast enough in height to keep up with competitors.

Analyzing your stands for young sapling and pole-sized trees that can be cut to develop sprouts and/or the presence of 100–200, 4-foot-tall advanced oak regeneration indicates the stand's ability to successfully maintain oak in a regenerating age class. If stands on medium- and high-productivity sites do not have these sources of reproduction, they will unlikely regenerate significant numbers of oaks after a typical regeneration treatment or harvest. If this is the situation, then silvicultural treatments must be used to build adequate numbers of appropriate sized advanced regeneration.

Management challenges

As indicated previously, oaks are intermediate in shade tolerance. This indicates an intermediate amount of sunlight must reach the forest floor for successful regeneration. Intermediate shade tolerance presents a challenge for management compared to species that are shade intolerant or shade tolerant. It is relatively easy to create full or nearly full sunlight from a regeneration treatment or harvest, thus setting the stage for the rapid growth of shade-intolerant species. Likewise, it is relatively easy to maintain shaded conditions on the forest floor that is conducive for the survival of shade-tolerant species and their ultimate regeneration. However, developing an intermediate amount of shade for the development of regenerating oaks requires a challenging degree of control and operational finesse.

The key to naturally regenerate oak is having the appropriate numbers and sizes of oak reproduction present in advance of planning a regeneration harvest. To develop these pools of advance regeneration or stump sprouts on medium- to high-quality sites, foresters must stop the development of an understory and midstory that produces too much shade. Removal of this shade, allowing advance oak regeneration to develop, and reducing competition are key components to successful oak management and silvicultural treatments that can be developed to improve oak regeneration on productive sites. The challenge is to plan ahead to create the conditions necessary for oak development (see Chapters 13 and 14). Forestry planning well in advanced of a regeneration harvest and conducting the silvicultural treatments necessary to culture advance regeneration are possible, but it takes long-term commitment.

Further suggested reading

Abrams, M.D. 2003. Where has all the white oak gone? *Bioscience* 53(10): 927–938.

Crow, T. 1988. Reproductive mode and mechanisms for self-replacement of northern red oak (*Quercus rubra*)—A review. *Forest Science* 34(1): 19–40.

Johnson, P.S., Shifley, S.R., and Rogers, R. 2002. *The Ecology and Silviculture of Oaks.* CABI Publishing, Oxford, U.K., 503 pp.

Loftis, D.L. 1990. Predicting post-harvest performance of advance red oak reproduction in the southern Appalachians. *Forest Science* 36(4): 908–916.

Sander, I.L. 1972. *Size of Oak Advance Reproduction: Key to Growth Following Harvest Cutting,* USDA Forest Service Research Paper NC-79. USDA, St. Paul, MN, 6 pp.

Stringer, J. 2006. Oak shelterwood: A technique to improve oak regeneration, Professional Forestry Note #3. Southern Regional Extension Forestry SREF FM 05 and University of Kentucky FOR-100, Lexington, KY, 7 pp.

section two

Silviculture:
What is in the tool box?

chapter seven

Regenerating oak stands the "natural" way

Callie Jo Schweitzer, Greg Janzen, and Dan Dey

Contents

Disturbance happens.. 75
Can we control disturbance to get what we want?..................................... 76
The clearcut method 77
The shelterwood method 78
Uneven-aged management... 82
Conclusion 83
Further suggested reading... 84

Disturbance happens

In forest stands, there is always a background of growth, senescence, and death of individual trees. There also is always some type of disturbance, from simple and constant events, such as wind, to more intensive and dynamic events, like a timber harvest. Hardwood stands are composed of multiple species, and these species respond differently to disturbance, depending on the age and size of each individual. So, when a stand is disturbed, you must consider not only the species but other factors as well. Every time a hardwood stand is disturbed (either via commercial harvest or other intermediate treatment, such as thinning or burning), there is an impact on the next stand (the regeneration). Disturbance must be coupled with what we know about regeneration response in order to accomplish a goal of regenerating oaks.

Two main characteristics drive the response of regeneration to disturbance—the species of regeneration and its shade tolerance (see Chapters 5 and 6). Some regeneration comes from new seedlings that develop after disturbance, typically light-seeded species, such as ash and yellow poplar. These species thrive under open conditions with plenty of sunlight (they are not tolerant of shade).

75

Other species, such as oaks, are considered advance-growth dependent, which means they must persist through the disturbance. As mentioned in Chapters 5 and 6, oaks are considered intermediate in shade tolerance, an inference to their ability to hang around in the understory (in the shade) and then respond to an increase in sunlight (more open conditions following harvesting, for example). The ability of oaks to survive through disturbance depends on the size of the regeneration and its growth rate. The probability of a small oak stem surviving and thriving after a disturbance increases with its stem size. The more oak stems you have before disturbance, and the bigger those oak stems are, the greater chance you have of maintaining oaks in your stands.

Hardwood trees also sprout when they are cut. We know the probability of having a stump sprout that will be competitive after cutting depends on the size and age of the cut tree and site productivity. Smaller, younger trees on less productive sites produce more stump sprouts. Large sawtimber-size trees will not produce stump sprouts that contribute significantly to oak regeneration in the next stand.

Can we control disturbance to get what we want?

If oaks are in your stands and they are producing acorns, you do not have a regeneration problem. What you may have, however, is a problem developing those oak seedlings into a competitive position. Many mast years are followed by a flush of small oak seedlings. If you do nothing to increase light to those seedlings, they will remain in the understory for several years, attaining a height of about a foot. After 10 years or so, they will die. This cycle is common without disturbance. If shade is the culprit, then the amount of sunlight hitting those small seedlings needs to be increased to promote growth. The catch is, in multiple species stands, when light is increased, everything responds. The race for dominance under these conditions is often won by shade-intolerant species—exactly the species competing with oak. So, how do we allow enough light to promote oak into a competitive position, but not open the stand so much that competitors take off?

A management prescription is stand specific. You must prescribe practices according to current conditions influenced by uncontrollable factors, such as past stand history and site conditions. Too often, we make broad assumptions about the condition of stands when prescribing management. For example, it is often assumed that stands are essentially the same age, also referred to as *even-aged*, though most stands have had partial harvests in the relatively recent past that resulted in older trees being left, while younger trees established. In reality, most hardwood stands contain trees in several age classes.

The clearcut method

Clearcutting can be a successful regeneration method in upland hard-wood forests. It involves removing all trees in one harvest treatment, providing full growing space for new trees to grow, whether planted or arising from seed or coppice (sprouts). This results in a new stand of trees essentially the same age, or even-aged. If implemented as such, this is correctly termed a "silvicultural" clearcut. In practice, loggers often leave small-diameter trees (not commercial), poor-formed trees (low grade), and/or species with little economic value and still call it a "clearcut." This use of the term is inaccurate and causes confusion. It is also not aesthetically pleasing to most people (if in view of roads) and degrades the regenerating stand significantly, depending on how many trees are left standing, their age, and species. When undesirable and poor-formed trees are left standing after harvest, they compete with the regenerating trees and the regeneration method is no longer a "silvicultural" clearcut, but often called a "commercial" clearcut.

Prior to clearcutting, there must be adequate regeneration of desired species. If the stand is ready to regenerate, it may be economically and biologically sound to remove the current stand. On less productive sites *where competition from species such as red maple and yellow poplar is restricted,* oak can successfully regenerate following clearcutting. On more produc-tive sites, *if adequate oak regeneration exists,* clearcutting releases these trees; however, it also releases competition. On productive sites, clearcutting is not the best method for regenerating oaks. Techniques that allow a slight increase of sunlight into the forest floor to stimulate oak regeneration, but not "release" competitor species, are needed. Once the oak seedlings have reached a height of 4–6 feet, a harvest should be considered to release the advance oak regeneration.

Economics often dictate the regeneration method used. If you need as much income from the stand as possible immediately, you may maxi-mize income by removing all trees in the stand, regardless of regenera-tion status. However, that decision comes with the risk of not regenerating species you favor, especially oaks on good sites. When clearcutting is con-ducted on productive sites, you must lower your expectations of having oak in the next stand. There may be a few oaks that survive the competi-tion, but the proportion of oak will be reduced considerably. If you have concerns about losing value or merchantability by leaving a portion of the trees in the stand (as would be done with a shelterwood harvest, see The shelterwood method section), that risk is removed by clearcutting.

You can enhance the probability of releasing regenerating oak from competition on good sites by identifying young, understory oaks and treat-ing adjacent competing stems with an herbicide treatment. This release must be done while the oak is still competitive, before it is overtopped by competing vegetation. The more oak seedlings you release and the larger those oaks are, the chance you will have oak in the next stand increases. It is important to realize this treatment comes at a cost, in both money (herbicide and labor) and time.

Clearcutting may be used to help meet objectives related to wildlife. When all trees in a stand are cut, maximum sunlight is allowed to reach the forest floor. This stimulates an immediate vegetation response, which temporarily provides habitat for a number of species that use early succession and young forests. Increased forage, browse, and soft mast are available for approximately 7 years without further disturbance until regenerating stems reduce the amount of sunlight reaching the forest floor to preharvest levels. The wildlife species that use developing young stands change as the vegetation composition and structure change. Clearcutting a hardwood stand sets back succession, but the site is dominated once again by young trees within 2 years as seedlings and sprouts reclaim the site.

Obviously, when a stand is clearcut, mast production from trees is limited for 30–40 years until the regenerating trees begin producing mast. When wildlife is an objective, one way to ameliorate this situation is to implement a clearcut with "reserves." By leaving reserves, that is, five or six mature mast-bearing trees with good form in the stand, some mast production is retained while the new stand regenerates. These trees can be harvested when the regeneration reaches mast-bearing age or they can be left and harvested when the regenerating stand is harvested.

From an economic standpoint, clearcutting with reserves makes little sense, especially when you consider the risk of wind damage to the trees retained. From an oak regeneration perspective, because oak is advance-growth dependent, leaving a few scattered oak trees postharvest will not contribute to oak regeneration. Any germinated acorns from those trees will not compete with the young forest that is already developing. Clearcutting with reserves should be considered when it is desirable to provide mast within dense cover (regenerating trees). This can be especially important for species such as ruffed grouse, white-tailed deer, and black bear.

The shelterwood method

The shelterwood method involves harvesting the stand through a series of partial cuts over several years but maintains an even-aged stand. The intention is to manipulate light and growing space so desired species (e.g., oaks) are favored over others and increase the probability of desired species dominating the next stand. Shelterwood harvesting can be used to successfully regenerate oak. However, advance oak regeneration must be present and you must consider the overstory species and the economics driving the harvests. The goal is to open up the stand so small oak regeneration can grow into a more competitive position, while sheltered by the remaining or residual trees. Once the oak is in a competitive position, residual trees are removed.

Residual trees (the "shelterwood") provide shade and protection for the regenerating trees below. In conifer systems, the shelterwood also provide a source of seed. In oak systems, the seed (acorns) have already fallen and germinated before the first shelterwood cut is implemented. As the regeneration develops, the shelterwood is removed through stages, usually in one or two cuts.

When managing for oaks, the purpose of the first cut is different than when managing conifer systems. When managing for oaks, the first cut may actually be a *thinning from below*, which removes stems from the midstory as well as some lower crown-class trees from the overstory (this may be 15%–25% of the stand's total basal area). The most important consideration when managing oaks is that oak seedlings (regeneration) are already present in sufficient numbers when the first cut is implemented. This is critical. The first cut allows only a relatively small amount of light into the stand to stimulate the oak seedlings and allows them to grow into a more competitive position (usually 4–6 feet tall). In many situations, the first cut is not a harvest. Sometimes, it is not even a cut. On some sites, sufficient sunlight is allowed to enter the stand by treating undesirable midstory species with an herbicide treatment (e.g., hack-and-squirt). It is important to realize this process is labor-intensive, relatively expensive, and not conducive to all stands or landowners (see Chapter 10). You also must be aware of the shade tolerance of other species in the stand. If there are shade-tolerant species, such as the sugar maple and American beech, treat these species accordingly, or they too will respond to the increased light and growing space created when the midstory dies.

The increased light allowed from the first cut is ephemeral. Not only will the oak regeneration respond, but so will other regenerating species, as well as any remaining midstory species. Once the oak regeneration is in an advance position, additional sunlight is allowed to enter from a second cut, which is a harvest, usually implemented 5–10 years after the first cut (depending on the site). In some situations, the shelterwood can be completely removed in the second cut. However, according to the other species present and the vigor of the competition, it may be advisable to remove a portion of the shelterwood and still provide partial shade to allow the oak regeneration to continue to grow. This is especially important if species such as yellow poplar and red maple are prevalent and vigorously competing with the oak regeneration. If a third cut is implemented, this is usually 5–10 years after the second cut.

Another approach that has been used with some success is using fire to help control competition after the initial cut, before the shelterwood has been removed. If competition is significant, prescribed fire implemented during the growing season 2–3 years following the initial cut will top-kill a majority of the existing stems, including the oaks. However, because oaks typically develop a strong root system early in life, they are able to send up a strong sprout the following year, enabling them to better compete with other species, such as yellow poplar and red maple, that do not have as strong root reserves.

The impacts of the shelterwood method on wildlife are considerable. Following the first cut, the understory is stimulated, providing nesting cover for many forest songbirds, brooding cover for wild turkeys and ruffed grouse, and increased forage and soft mast for other species, including white-tailed deer, black bear, and raccoons. Following the second harvest, stem density in the midstory increases substantially. This provides cover for another suite of species, such as chestnut-sided warbler, eastern towhee, brown thrasher, and ruffed grouse. If some of the shelterwood remains, an added benefit is the mast available from those overstory trees, all within the cover of the developing regeneration. Along with those species that require dense developing cover, other species that nest and forage in overstory trees, such as tanagers, red-eyed and yellow-throated vireo, and black-throated blue warbler, may still be found in these stands. Thus, a diversity of wildlife species is found in stands that contain diverse vegetation structure and species composition.

If you are trying to manage property for wildlife species that benefit from acorns and dense cover, an option is an irregular shelterwood. This method allows a certain amount of the overstory (usually 20–50 square feet of basal area per acre) to remain for an extended period while the regeneration develops. The residual trees can be removed for increased timber value several years later or they can be retained in the stand until the next rotation, providing increased wildlife value by continuing to produce mast within the developing cover. Obviously, it is necessary to retain species with good mast value (e.g., oaks, persimmon, blackgum) and to not retain trees that may produce increased competition (such as yellow poplars, maples, ashes).

The shelterwood method involves multiple entries into a stand. Anytime you increase the number of harvest entries, you increase the risk and potential cost. One risk is damaging residual trees during the first or second harvest. A logging operation will impact residual trees; leaving fewer trees will decrease the number the logger has to dodge around, but you may not have enough volume for the final harvest. The residual trees must be able to sustain another harvest, so species composition and merchantable condition (size and grade) must be adequate. If it is not feasible to leave enough merchantable trees for the final harvest, it is best to allow the stand to grow and harvest at a later date.

Shelterwood harvesting that targets suppressed and overtopped trees may create conditions favorable for oak regeneration, but the amount of sheltering wood, or residual trees, varies with site productivity. On more productive sites, more trees must be retained because there is a greater probability for competitive species to take over. You should also harvest competitive species, such as the yellow poplars, ashes, and maples, and reduce some of the potential competitive seed source. There is a fine line between allowing enough light for oak while not encouraging competitors. Of course, a registered forester should help you with these decisions.

The influence of residual trees affects the growth rate and the species composition of the nearby reproduction. Reproduction of all species is taller with larger diameters the farther they are from residual trees. However, the rapid growth of intolerant species, such as the yellow poplar and red maple, allow them to dominate the regeneration. Some treatment to release desirable species to meet management objectives may be beneficial. The duration of sheltering is varied and depends on the site and stand development. An assessment of the oak regeneration is required to drive the decision to conduct a release of favorable trees or the final harvest.

The initial disturbance, or harvest, and the resulting residual conditions, influences regeneration. All species will respond to an increase in light. The response depends on species present, site conditions, and stand conditions. On productive sites, the regeneration cohort soon starts to compete with itself and the effect from the residual stand is diminished. Removing the residual stand while oak is still competitive increases the chances of oak becoming dominant in the stand.

On highly productive sites, a midstory herbicide treatment is often recommended, as openings created by a timber harvest may accelerate the growth of competing tree species beyond that of oak. However, the economics of this treatment my render it unfeasible. If the initial shelterwood treatment on productive sites is a partial harvest, it may be necessary to go into the resulting regeneration and use an herbicide treatment to release oak prior to the final harvest.

Do not confuse the shelterwood method with a diameter-limit harvest. What typically happens when using diameter limits to drive a harvest is the best trees are taken in the first harvest and only "junk" or marginally valuable trees remain. The residual stand is usually of much lower quality, includes a greater proportion of less desirable species, and has reduced merchantability. The final harvest is more difficult and often not accomplished. These retained low-quality trees influence the resultant stand until the next rotation. The regeneration is never truly released and is compromised. Diameter-limit cutting is not recommended for regenerating upland hardwood stands.

Foresters often assess the cubic volume of stemwood and stem density by measuring the cross-sectional area of trees 4.5 feet aboveground (see the sidebar on "DBH," Chapter 12). For example, a tree with a diameter of 13.5 inches has a cross-sectional area of 1 square foot; a tree with a diameter of 16.6 inches has 1.5 square feet; a tree with 19.5 inches diameter has 2 square feet. This cross-sectional area is referred to as *basal area*. By adding the basal area of each stem on any given acre, a measure of tree density can be determined. If there is 100 square feet of basal area on an acre (a reasonable figure for many mature oak forests in the Eastern United States), that can be comprised of many small-diameter stems, a smaller number of large-diameter stems, or some combination of small- and large-diameter stems. Basal area is fairly well correlated with the cross-sectional area of the crown. Thus, in most closed-canopy stands, if 40% of the basal area is removed, you may expect to allow approximately 40% sunlight into the forest floor.

Uneven-aged management

Uneven-aged management for sustaining oak stands has been marginally successful and is again dependent on the site and stand characteristics. The premise is to create three or more age classes within the stand with partial harvests over time. Each partial harvest results in a new regeneration age class. These harvests are meant to be sustainable over time. Problems include having enough merchantable trees to make repeated commercial harvests and creating conditions to allow regeneration of desirable species.

There are two common uneven-aged systems: single-tree selection and group selection. In single-tree selection, periodic harvesting of trees of all species in all diameter classes is performed. In reality, the infrastructure of maintaining roads and trails, and marking the timber is difficult. Desirable species (including oaks) do not regenerate or do not grow into competitive positions with single-tree selection. In hardwoods stands in the Eastern United States, single-tree selection favors shade-tolerant species, such as the sugar maple and American beech, not oaks.

The other uneven-aged system is group selection, where openings larger than the height of the largest trees in the stand (0.2–1.5 acres) are harvested, similar to small clearcuts, but within a given stand.

As with any other method, adequate advance oak regeneration must be present prior to creating these openings or oak will not be successful in the future stand. Conditions conducive to regenerating and recruiting oak into competitive positions are created in the perimeter of these gaps where sunlight is more intermediate. Again, consider the infrastructure, as well as maintaining and developing enough merchantable stems to sustain the harvests.

Group selection can provide diverse cover and food resources for wildlife. Not only are conditions improved for those species that would benefit from clearcutting and shelterwood harvesting, but also many forest interior birds, such as cerulean warblers, select these small openings for foraging. Group selection harvest units can be placed across a property in such a way that they provide "successional stepping stones" connecting larger units of young forested cover and/or areas of early succession.

Conclusion

The clearcut method of regenerating upland hardwood stands is appropriate where existing numbers of advance oak reproduction is adequate, which may occur on most xeric sites. The more common situation is a lack of advance oak reproduction, or numerous small seedlings. In this case, the shelterwood method is most useful, is flexible to tailor its application for existing stand conditions, and can be combined with measures to control tolerant midstory species.

The bottom line in regenerating oak is advance regeneration must be present before harvesting. Getting advance oak regeneration prior to harvest can be challenging, but can be accomplished through intensive methods, such as herbicide treatments or by treatments that are carefully timed, such as prescribed fire, and applied based on individual stand characteristics.

Not only must advance oak regeneration be present, it must also be of significant stature to be competitive; a common rule is greater than 0.25–0.5 inch basal diameter. It is important to release oak regeneration before you lose it. Lack of light limits the development of oak seedlings in forest understories. You can control light by reducing stand density. On high-quality sites, the recommendation is to do this gradually, over 10–20 years, first by removing the midstory, then by decreasing overstory density in one or more removal harvests. Control competing regeneration anytime it threatens to suppress oaks.

Think of regenerating a stand as a process of repeated disturbances to favor oak, not as a single event. Assessing the potential of a stand and delivering timely and adequate disturbances with patience can regenerate upland hardwood stands to favor oaks.

Further suggested reading

Burger, G. and Keyser, P. 2013. The shelterwood-burn technique for regenerating oaks, PB 1813. University of Tennessee Extension, Knoxville, TN.

Dey, D.C. and Parker, W.C. 1996. Regeneration of northern red oak (*Quercus rubra* L.) using shelterwood systems: Ecophysiology, silviculture and management recommendations. Ontario Forest Research Institute Forest Research Information Paper 126, Sault Ste. Marie, Canada.

Gingrich, S.F. 1967. Measuring and evaluating stocking and stand density in upland hardwood forests in the central states. *Forest Science* 13:38–53.

Johnson, P.S., Jacobs, R.D., Martin, A.J., and Godell, E.D. 1989. Regenerating Northern red oak: three successful case histories. *Northern Journal of Applied Forestry* 6:174–178.

Johnson, P.S., Shifley, S.R., and Rogers, R. 2002. *The Ecology and Silviculture of Oaks*. CABI Publishing, New York, 503 pp.

Jones, B.C. and Harper, C.A. 2007. Ruffed grouse (*Bonasa umbellus*) use of stands harvested via alternative regeneration techniques in the southern Appalachians. In *15th Central Hardwoods Forest Conference*, Knoxville, TN, GTR-SRS-101, pp. 373–382.

Lashley, M.A., Harper, C.A., Bates, G.E., and Keyser, P.D. 2011. Forage availability for white-tailed deer following silvicultural treatments. *Journal of Wildlife Management* 75:1467–1476.

Loftis, D.L. 1990. A shelterwood method for regenerating red oak in the southern Appalachians. *Forest Science* 36:908–916.

Miller, G.W., Kochenderfer, J.N., and Fekedulegn, D.B. 2006. Influence of individual reserve trees on nearby reproduction in two-aged Appalachian hardwood stands. *Forest Ecology and Management* 224:241–251.

Putnam, J-A., Furmival, G.M., and McKnight, J.S. 1960. *Management and Inventory of Southern Hardwoods, Agriculture Handbook*, No. 181. U.S. Department of Agriculture, Forest Service, Washington, DC, 102 pp.

Sander, I. 1971. Height growth of new oak sprouts depends on size of advance reproduction. *Journal of Forestry* 69:809–811.

Smith, D.M. 1986. *The Practice of Silviculture*, 8th edn. John Wiley & Sons, New York, 527 pp.

Weigel, D.R. and Johnson, P.S. 1998. Stump sprouting probabilities for southern Indiana oaks. Technical Brief No. 7, U.S. Department of Agriculture, Forest Service, North Central Forest Experiment Station, St. Paul, MN, p. 7.

chapter eight

Artificial regeneration

David S. Buckley and Victor L. Ford

Contents

Reasons for using artificial regeneration 85
Artificial regeneration scenarios 86
Matching species and sites... 87
Seed sources 89
Direct seeding 90
Planting nursery seedlings 93
Seedling protection 95
Fertilization and weed control 96
Artificial regeneration investments versus goals 97
Further suggested reading... 97

Artificial oak regeneration involves planting acorns or nursery seedlings, in contrast to natural regeneration, which originates from naturally dispersed acorns or sprouts. Depending on the circumstances, planted seeds or seedlings may be the sole source of oak regeneration, or will be used in combination with natural oak regeneration.

Reasons for using artificial regeneration

An important incentive for implementing artificial regeneration is a complete lack of natural oak regeneration sources in the vicinity, namely, seed-producing mature trees, established seedlings and saplings, and oak stems capable of producing sprouts after cutting. Sprouts originating from dormant buds on oak stumps and roots serve as a source of natural regeneration if seed sources and seedlings are lacking, but sprouting tends to diminish with oak stem size and age. Thus, sprouting is not really an option in stands dominated by large, old oak stems.

Reasons for supplementing natural regeneration with artificial regeneration include the desire to (1) increase the diversity of oak species present on a site (termed enrichment planting), (2) reinforce the numbers of regenerating oak stems (termed a reinforcement planting), or (3) introduce particular genetic lines to increase overall stand quality. Also, if creating or improving wildlife habitat is an objective, such as providing acorns as a food source, artificial regeneration can be used in areas lacking natural regeneration sources, such as legacy mine sites undergoing reclamation. Many characteristics that affect the quality and value of oak timber are inherited, and commercially important oak species are a focus in several current tree improvement programs. If certain genetic families having enhanced growth, timber quality, or acorn production are identified and selected, artificial regeneration will be the vehicle for introducing or reintroducing these desired characteristics into future oak forests.

Artificial regeneration scenarios

Acorns and nursery seedlings can be planted beneath mature canopy trees or in open areas. Planting seed or seedlings beneath mature overstory trees is referred to as *underplanting*. Underplanting acorns or seedlings in established forests tends to be more experimental and less common than planting in open areas. As described in Chapters 6 and 7, the intensity of competition between young oak stems and mature overstory trees must be evaluated and addressed. To avoid damage, the most logical time to underplant acorns and oak nursery seedlings is immediately after the removal of overstory trees. It is also best to plant after prescribed fire or herbicide treatments are implemented to control understory competitors. Subsequent treatment of weeds and other competitors with herbicide and mechanical methods is possible, but additional steps must be taken to protect planted oaks. Planting oaks after harvesting is more labor-intensive than planting prior to cutting if treetops and slash are abundant, but tree felling and skidding can wipe out many planted seedlings, and substantially disturb planted acorns.

Several opportunities to plant acorns and oak seedlings in the open are provided under reforestation (replanting forests soon after deforestation), afforestation (planting forests on sites with a different prior land use, such as agriculture or a different vegetation type, such as grassland), and species conversion scenarios. Afforestation efforts are currently underway to restore bottomland hardwood forests on substantial acreage cleared for farming. The sidebar in this chapter highlights bottomland afforestation efforts in West Tennessee. Oak species are a desirable and successful component of hardwood mixtures planted on reclaimed mine sites in the Eastern United States, and researchers working under the Appalachian Regional Reforestation Initiative have developed a Forestry Reclamation Approach that, when properly applied, enhances oak seedling survival and future forest productivity. An interesting example of afforestation is located in the Nebraska National Forests and Grasslands, in which the largest human-established forest was planted in former grassland in 1902.

Species conversion scenarios involving clearcutting established forest stands, herbicide treatment of any hardwood stumps, and subsequent planting of oak are also possible, but often limited to experimental trials conducted by managers and researchers. Conversion of poor-quality oak stands and other hardwoods to planted pines are far more common, particularly on sites that are intermediate in productivity.

Matching species and sites

An improper match between the site and species planted is one of two major reasons underlying the failure of oak plantings. Just because a landowner or manager wants a certain species to grow in a particular area, it does not mean it is feasible. An example occurred in oak plantings just south of the confluence of the Mississippi and Ohio rivers. The median soil pH in this area is about 8.0 and not conducive to oak plantings. Tests of several oak species on these soils resulted in planting failures, and only overcup oak survived in very poorly drained soils. If providing food and cover for wildlife is part of the objectives, then planting species that are native to the area and appropriate for the site (ridgetop versus bottomland) will provide the greatest wildlife benefit. Matching the species to the site is a tried and true principle in silviculture as well as agriculture. As discussed in Chapters 3 and 5, oak species differ in many characteristics and are not equally interchangeable on many types of sites. This is especially true at the very dry (e.g., ridgetops, upper slopes) and very wet (e.g., bottomlands) ends of the soil moisture spectrum. Dramatic changes in the suitability of sites for different bottomland oak species can occur with very slight changes in elevation in bottomlands. You should consult the section on silvics (i.e., adaptations to soils and soil factors, light conditions, competition, and disturbances like fire and flooding) in Chapter 5 when matching sites with oak species. The current online version of the USDA Forest Service silvics manual (www.na.fs.fed.us/spfo/pubs/silvics_manual/table_of_contents.htm) is another excellent source of information on site requirements for different oak species. Selection of appropriate oak species to be used in artificial regeneration depends on identifying site conditions affected by factors such as aspect (north-facing slopes, south-facing slopes, etc.), elevation, soils and any competitors present, and selecting oak species adapted to the particular conditions on the site. A useful method of verifying appropriate site-species matches is to identify and tally the particular oak species present in areas with similar site factors (aspect, slope, elevation, etc.) in the immediate vicinity of the planting site. A proven method to properly interpret site characteristics and select appropriate oak species is to consult a forester. A mixture of oak species is always desirable for increasing the likelihood of acorn production in any given year, but site conditions may limit the possibilities to only one or two oak species.

RESTORING BOTTOMLAND HARDWOODS WITH HIGH-QUALITY OAK SEEDLINGS IN WEST TENNESSEE

Jason S. Maxedon

Wetlands Wildlife Forester, Tennessee Wildlife Resources Agency

Efforts to restore bottomland hardwoods through artificial regeneration have become increasingly common due to decreased profitability of agricultural production and persistent flooding of marginal farmland in river bottoms. Over the past decade, the Tennessee Wildlife Resources Agency has been acquiring former bottomland agricultural sites in West Tennessee for conversion to wildlife habitat.

A planting program using high-quality oak seedlings grown from local seed sources was initiated in 2000 to provide a future source of hard mast in these areas. Acorns have been collected from local mother trees representing nine oak species: willow oak, water oak, cherrybark oak, pin oak, Nuttall oak, overcup oak, swamp chestnut oak, bur oak, and Shumard oak.

Seedlings of these species were initially produced at the Georgia Forestry Commission's Flint River Nursery using special fertilization and irrigation protocols developed by Dr. Paul Kormanik (U.S. Forest Service, retired) and others, which allow seedlings to attain maximum potential size in a single growing season. Similar protocols have been subsequently adopted by the Tennessee state tree nurseries, and these nurseries serve as the current source of high-quality seedlings for bottomland hardwood restoration.

High-quality, bottomland oak seedlings have been established on more than 5000 acres spread over more than 18 sites, which range from 3 to 760 acres in size. Species were initially matched to site by visual inspection, which evolved into species/site matching by elevation. Survival and growth have been evaluated through sample plots at each planting site.

Survival has generally ranged from 60% to 100% with over half the sites having 70% or better survival. As a result, the plantings can be considered an overall success. Growth averages approximately 2 feet per year after the third year of planting, with phenomenal growth occurring in some areas (30 feet in 8 years!).

> Approximate planting costs include the following:
> Restoration of hydrology (where needed) = $200.00/acre
> Site preparation (mechanical or chemical) = $30.00/acre
> Contour survey and mapping = $12.00/acre
> Currently planting 435 trees/acre on a 10 × 10 foot spacing (65% oak and 35% other mast producing species)
> Seedling cost = 283 high-quality oaks/acre @ $0.40 ea. = $113.20/acre
> Seedling cost = 152 other species/acre @ $0.25 ea. = $38.00/acre
> Planting labor = $85.00/acre
> Mechanical or chemical postplanting maintenance (where needed) = $20.00/acre
> The average cost to restore a site has been running around $265.00/acre.
> Site prep and postplanting maintenance are not always performed.

The larger size of high-quality seedlings translates into greater investments in seedling production and planting labor, but this tends to be balanced by greater survival, a reduced need for competition control, an increased ability of seedlings to tolerate deer browsing, and phenomenal growth when species are matched with optimum sites.

Seed sources

Selection of seed sources is an important second step in artificial regeneration. Future quality of the oaks planted is often closely linked to the quality of trees used as seed sources. Local environmental conditions and history influence the quality of a particular mother tree, but high-quality genetic material is more likely to come from a high-quality seed tree than one that is low in quality. Another characteristic to consider when selecting seed sources is the degree of adaptation of the seed source to conditions on-site. Regional varieties with different tolerances and degrees of adaptation to factors such as temperature, rainfall, and soils occur within many hardwood species. Genetic differences across shorter distances (say, from one end of a state to another, or even from one county to another) can also be significant if changes in soils, elevation, temperature, and moisture are significant. Substantial and abrupt changes in soil types or elevation may be responsible for significant genetic differences over short distances. A useful rule is to collect or purchase acorns produced by high-quality trees in locations that are geographically and ecologically similar and located as close as possible to the planting site.

Depending on the supplier, oak nursery seedlings may or may not have been grown from acorns produced by high-quality seed sources and may have been grown from seed sources collected hundreds of miles from the intended planting site. A reputable nursery will divulge information on the location and quality of seed sources used to produce seedlings, and some will work with consumers on growing seedlings from seed collected by the consumer. There is an increasing trend toward custom production of seedlings from strictly local seed sources for planting in preserves, state parks, and national parks, in which a primary management goal is conservation of native biodiversity, including native genotypes. Such exacting seedling production methods are not always possible or practical for a variety of reasons, so seed sources from locations as close as possible to the planting site often represent the next best strategy. Interestingly, if there are no significant changes in elevation, better growth can be achieved by planting acorns 200 miles to the north of the collection site. Doing the opposite (planting acorns 200 miles to the South of the collection site) generally results in less growth due to those ecotypes breaking bud later than seedlings grown from local seed.

Direct seeding

Sowing tree seeds in or on the soil is called *direct seeding*. Direct seeding is initially less expensive than planting nursery seedlings because direct seeding does not involve the costs associated with producing, lifting, shipping, storing, and planting nursery seedlings. Acorns are also logistically easier to handle and plant than nursery seedlings. Acorns can be planted with hand tools, and mechanized acorn planters are designed for larger areas. Finally, distinct advantages of direct seeding are that seedlings will develop root, shoot, and leaf characteristics that are best adapted to local conditions, and that there are no problems with transplant shock.

You can collect acorns from high-quality seed sources near planting sites to reduce costs associated with direct seeding and ensure establishment of locally adapted trees. To maximize viability and reduce chances of drying out, fungal infection, and insect infestation, acorns should be collected as soon as possible after being shed by the parent tree. Collecting acorns from tarps spread beneath seed trees, instead of allowing acorns to contact leaf litter and soil, reduces problems with fungi and insects.

Acorn weevils significantly reduce seed viability by depositing eggs in acorns before they are dispersed from parent trees. Acorn weevil larvae can be present in a high percentage of the acorns collected. Infested acorns containing weevil larvae can be separated from sound acorns by dropping them in water. Infested acorns tend to float, and more sound acorns with a higher probability of successful germination tend to sink. There are a few exceptions in which a few infested acorns will sink, but float testing is the most efficient, nondestructive method for testing acorns.

Acorns from species in the white oak group (e.g., white oak, chestnut oak, bur oak) will germinate in the fall just after dispersal from the parent tree, whereas acorns from oaks in the red oak group (e.g., northern red oak, black oak, and pin oak) need a period of winter temperatures to break dormancy, and therefore germinate in the spring. Acorns from the white oak group should be planted in the fall to avoid damage to the tiny shoots and roots that rapidly emerge. Acorns from the red oak group can be planted in either the fall or spring, but are best planted in the fall to provide the winter temperatures needed to break dormancy. When planting acorns from the red oak group in the spring, you should refrigerate them over the winter in plastic bags with a few drops of water and a teaspoon of ethyl mercaptan fungicide. Bags can then be stored over the winter in a refrigerator or walk-in cooler with the temperature maintained a little above freezing (approximately 40°F).

Despite the appeal of direct seeding, there are several disadvantages over planting nursery seedlings. A basic difference between acorns and nursery seedlings is that acorns can be sorted by weight and size to some extent prior to sowing, but unlike nursery seedlings, they cannot be graded on the basis of many characteristics related to competitive ability and quality, such as stem size, form, and root system development.

Mammals and birds can remove significant numbers of acorns after sowing. To be sure, a few of these acorns simply may be moved around and replanted by small mammals and birds, such as blue jays, but large percentages can also be consumed and destroyed, leading to patchiness or complete failure of acorn plantings. The degree of oak seed predation also depends on local populations of deer, small mammals, and birds and can depend on the availability of cover in the vicinity of planted acorns. In a Michigan study, the percentage of seed planting locations with acorns removed by rodents was greater in forest areas with high canopy cover than in adjacent clearcuts.

Apart from potential losses to wildlife, another disadvantage of direct seeding is uncertainty regarding germination. Stretches of drought and extreme temperatures can reduce germination and early survival, and very small-scale differences between sites with planted acorns can lead to patchiness and a lack of control over spacing. Seedbeds vary considerably from one part of a forest stand to another, and the likelihood of obtaining one successful oak seedling per planting location or seed spot is often very small. A final disadvantage of direct seeding is the small size of first-year seedlings substantially reduces their growth rates, competitive ability, and likelihood of survival. Several years may be needed to develop the sizes of advanced oak reproduction required to successfully compete with other tree species following various types of harvest, and other silvicultural treatments, such as prescribed burning. Failures resulting from acorn predation, poor germination, or poor early survival are not always the case, however, as evidenced by comparable levels of success achieved with direct seeding and planting nursery seedlings on bottomland sites in Louisiana and Mississippi.

Direct seeding oaks can be accomplished by randomly broadcasting acorns across the forest floor or soil by hand or with a broadcast spreader, or by planting in the soil to 4 inches deep. Acorns can be planted by hand with a trowel, sharpened stick, or hoe, or with a mechanized acorn planter. Dibbles also have been used but tend to result in planting depths deeper than 4 inches. Broadcast sowing can be advantageous in underplanting scenarios when equipment operation is hindered by the presence of overstory trees and is the least labor- and equipment-intensive sowing technique. Distinct downsides to broadcast sowing are the increased likelihood of acorns drying out and not germinating, or being moved or eaten by mammals and birds. These losses can be offset by increasing the number of acorns sown per acre. Planting acorns in the soil by hand or machine has the advantages of putting acorns firmly in contact with moist mineral soil and reducing losses to wildlife, but represents a greater investment than broadcast sowing. Agricultural seeders have been modified to plant acorns for use in open areas and fields. Conventional planters require considerable soil preparation to allow the planter to work. No-till or minimal-till drills do not require soil preparation much beyond debris removal. This technique works best with small red oak acorns. Acorns larger than ½ inch in diameter are difficult to machine plant. White oaks generally develop seedling roots soon after falling from the parent tree, and these seedling roots have a tendency to break during planting with a machine. The appropriate acorn sowing rate depends on the number of acorns available, the desired future number of oak stems per acre, and the likely effects of seed predators and adverse site conditions. Rates of 700–1000 acorns per acre can achieve desired results on bottomland sites, 1000–1500 acorns per acre are commonly planted on most sites, greater densities (1200–1500 acorns per acre) can help increase the chances of success on sites having more barriers to successful germination and survival, and rates as high as 2400 acorns per acre can ensure success on sites with very heavy acorn predation.

Planting nursery seedlings

Oak nursery seedlings vary widely in size and quality. In general, larger seedlings with better-developed tops and root systems have a greater ability to compete for light with herbaceous and woody vegetation, and to survive damage from browsing, defoliating insects, fire, and late spring frost than smaller seedlings. Greater aboveground seedling heights place leaves in a more competitive position in relation to other understory vegetation and hasten the growth of terminal buds beyond the reach of deer. Larger root systems and stems translate into greater amounts of stored energy available during establishment, and to replace leaves and branches lost to damage. Ideally, the diameter of taproots at the soil line (also called the *root collar*) would be larger than ½ inch and the top should have well-developed buds. Height growth after outplanting has been correlated with the number of coarse (as thick or thicker than pencil lead) lateral roots attached to the taproot.

Despite the advantages of large, high-quality seedlings discussed earlier, major trade-offs associated with large seedlings are increased costs per seedling for production, transport, and planting. There are also practical limits on seedling size that can be transported and planted properly. A full spectrum of seedling sizes and quality is available from 4- to 5-foot-tall seedlings transported to the planting site in pots (called *containerized seedlings*) to seedlings less than a foot tall lifted from the soil (called *bare-root seedlings*). In addition to the containerized and bare-root classifications, oak seedlings are often categorized by age, number of years spent in the nursery bed where they were sown, and number of years spent in a transplant bed. A two-digit code is assigned to seedlings, with the first number corresponding to years of growth in the bed where they were sown and the second number corresponding to the number of years of growth in a second transplant bed. A 1-0 seedling, for example, is 1 year old, grew a year in the bed where it was sown, and was not transplanted. A 1-1 seedling and a 2-0 seedling are both 2 years old, but the 1-1 seedling was transplanted after a year, whereas the 2-0 seedling remained in the bed in which it was sown for 2 years. Although 1-1 and 2-0 seedlings are older than 1-0 seedlings, they are not necessarily larger. Variables such as oak species, genetic family, sowing density, fertilization, irrigation, and geographic location of the nursery have a strong impact on seedling size and other characteristics. A batch of 2-0 bare-root seedlings grown in the upper Lakes States may be less than a foot tall, whereas 1-0 bare-root seedlings grown in the South may be 4–5 feet tall.

The length and especially the width of root systems are often the most limiting factors affecting the feasibility of planting very large oak seedlings. A root system that is simply too large often results in trees planted in a shallow hole, especially on rocky planting sites such as those created during mine reclamation. Upper portions of the root system are either killed by exposure to the sun, air, and wind in planting holes that are too shallow, or absorb less soil moisture than those planted deeper in the soil profile. Twisted "J" roots that grow toward the soil surface instead of toward the deeper, moister soil horizons may also develop if large seedling root systems are forced into holes that are too narrow or shallow.

Root pruning can make large seedlings easier to plant but can decrease their root growth potential. Reduced root growth can translate into less seedling growth and lower survival. The growth of leaves and stems depends on having an adequate root system to supply water and nutrients, and likewise, root systems require adequate numbers of photosynthesizing leaves to supply energy for root growth and maintenance. Sustaining a reasonable balance between roots and tops is important for overall seedling growth and survival. As a result, extensive pruning of either the roots or tops of bare-root seedlings should be avoided.

Seedlings are stored in nursery bags in a cold room after lifting from nursery beds and need to remain between 33°F and 40°F until planted. Exposure of seedlings to sun and higher temperatures over a number of days, which can occur when nursery bags are left in the backs of unrefrigerated trucks during planting, can lead to substantial losses of seedling vigor. Bare-root nursery seedlings are lifted in the winter in the South and in early spring in the North, and should be planted as soon as possible after lifting and bagging. Cold weather during and after planting ensures bud break and expansion of leaves will not occur before seedlings have the opportunity to begin at least a small amount of root growth on the planting site. Oak seedlings planted too late in the spring will break bud immediately, and the inability of the root system to supply sufficient moisture often results in partial dieback of leaves and terminal portions of the stem, or death of the whole seedling. Seedlings are fragile and should be handled as such. Bare roots should not be exposed for more than 20 seconds to the sun, wind, and air during planting. Planting bags filled with wet peat moss, wet moss collected from the site, or wet paper are designed to protect trees seedling root systems and should be used to transport seedlings from nursery bags to planting locations. The best planting days are overcast with little wind, high humidity, or even light precipitation.

Trees can be planted with shovels, dibbles, or mechanized tree planters. Planting seedlings a little deeper on the planting site than in the nursery bed will ensure seedlings are set tightly in the soil and that soil contact is maximized. Backfilled soil often settles or erodes after planting, which can result in exposure of the top portion of the root system. In underplanting scenarios in which there are overstory trees, the forest floor and fine root mat that occurs near the soil surface can be undercut in a circular pattern, lifted, and cut in half with a shovel and set aside, while the rest of the planting hole is excavated. If the two halves of this forest floor and root mat section are replaced on either side of the planted seedling stem and firmed into place, they will act as mulch and reduce the evaporation of water from the soil surrounding the seedling. Use of this method resulted in 99.9% survival of 1152 bare root northern red oak seedlings planted on sandy soils in a Michigan oak study. Larger hardwood dibbles are generally required to plant oaks, which tend to have larger root systems than many other tree species. Shovels and sharpshooters (tile spades) are more effective for seedlings with larger root systems, but require more time per planted seedling than dibbles. Soil augers have been used to excavate planting holes, but there is a tendency to slicken and compact the sides of the holes in clay soils under wet conditions, which can retard root penetration into the surrounding soil. The roots never leave the hole in severe cases as if the seedlings were planted in pots buried in the soil.

Benefits of containerized seedlings over bare-root seedlings are longer planting season, increased growth, and increased survival. Containerized seedlings, however, are usually more difficult to transport and handle and are more expensive. They may arrive in the container or already pulled from the container in seedling bags. If they are in seedling bags, they should be treated as bare-root seedlings. Dibbles, shovels, and mechanical planters all have been used to plant containerized seedlings. One of the greatest advantages of containerized seedlings is the opportunity they provide for fall planting. As long as they are planted in an area that does not have the danger of frost heaving (loosening of the soil and roots from freezing and thawing), containerized seedlings become better established if planted during the fall and will grow and survive better during the first year.

Tree spacing and stocking are always sources of controversy among foresters. Factors such as diameter growth, potential survival, size desired for first thinning, and amount of self-pruning must be weighed to determine proper spacing. Most foresters agree that 700 trees per acre represents a maximum stocking, and at least 350 trees per acre are needed for a minimum stocking. If other desirable species, such as ash, seed in naturally, oak stocking can be further reduced. Optimal oak stocking is about 450 trees per acre.

Seedling protection

Oak seedlings are susceptible to a wide variety of agents that affect growth and survival. There are numerous leaf-eating insects that can affect oak, but most may be controlled with registered insecticides. Piercing sucking insects, such as scales, may be more difficult to control, however, and may require a registered systemic insecticide. The local university Extension Service can help you identify the best control agent. As a rule, oak stands are not treated until visible damage/symptoms indicate a loss of vigor. Preventative treatments are generally not effective in terms of control or cost.

Animal damage from rodents, such as voles, is common when herbaceous vegetation or other forms of cover are present around seedling stems. These animals girdle and can kill young oak seedlings. The best control for voles in these situations is effective herbaceous weed control to remove escape cover, which exposes the animals to predators. Rocks and boulders on reclaimed mine sites provide similar cover for rodents, as well as deep snow in northern areas.

One of the most significant sources of damage to oak plantings is browsing where deer are overabundant. Deer may browse planted seedlings year round, but winter damage is most noticeable. Newly planted nursery seedlings with high mineral nutrition are often more likely to be browsed than natural seedlings on sites, and fertilized seedlings seem to be especially palatable. The fastest growing, most successful seedlings seem to have the greatest risk of being browsed. Impacts of deer browsing can vary with forest structure. In a Michigan study, the incidence of deer browsing was greater in small clearcuts than in shelterwoods and uncut areas. Repellents, such as hair or soap, are only effective for a couple of weeks until they lose their smell, and commercial repellants, though water fast, are not effective on new growth. Repellents sprayed in the winter may prevent browsing, but the situation can be hopeless during the growing season. The overall effectiveness of repellents applied on an individual oak stem basis may depend on the availability of alternative food sources. Again, this is problematic where deer are overabundant. Chapter 17 describes problems associated with overabundant deer herds in detail. Fences are effective, but can be cost prohibitive except for small plantings up to a couple of acres. Debris, such as treetops left after harvesting, can help protect seedlings, but hungry deer may yet find them. Tree shelters can be effective in protecting oak seedlings from deer. Tree shelters are translucent solid plastic or mesh tubes that provide protection by encasing the growing seedling. Solid tubes promote height growth and offer some protection from herbicide sprays. Most tree shelters are about four feet tall and have to be removed once the seedling is taller than the tube. However, some are biodegradable.

Fertilization and weed control

Fertilization can help oak seedlings establish if a particular nutrient is limiting. In general, however, fertilization is detrimental. Weedy competitors are fertilized and stimulated along with the oak seedlings, and the increased weed competition often overwhelms the seedlings. It is more effective to invest in better weed control than to fertilize. One of the reasons seedlings in recent cutovers do not need fertilizing is the Assart Effect—the temporary increase in nutrient availability that occurs after harvest and site preparation from decomposition of harvest residue, which typically exceeds nutrient uptake by newly planted seedlings. Demand generally outstrips supply about the time of crown closure. Fertilization is generally more effective if delayed; fertilization at planting rarely pays.

A second major reason that oak plantings fail is either improper or no weed control. This is particularly true on open sites with abundant light and belowground resources, such as former agricultural fields in bottomlands undergoing afforestation. Increased natural fertility and productivity of sites leads to increased competition between planted oaks and weeds, just as fertilization tends to increase the intensity of competition. Weeds prevent establishment of oak root systems and compete with seedlings for belowground resources and sunlight. Whether the competitors are tree saplings of other species, shrubs, or herbaceous weeds and grasses, the result is the same: poorly stocked stands with low vigor. The root system of oak seedlings must become established before significant height growth takes place. Oak seedlings surviving heavy competition often take considerably longer to begin the height growth phase. A number of herbicides and herbicide application methods are available for controlling weeds, as well as mechanical site preparation methods, such as disking (see Chapter 10).

Artificial regeneration investments versus goals

As outlined earlier, a rather wide variety of artificial regeneration options exist for landowners and managers. Foresters like to quip that oak regeneration is a sure thing on any site as long as sufficient quantities of funds are expended. The most appropriate artificial regeneration methods depend on the level of natural oak regeneration present, what particular barriers to regeneration are most important on the site, the financial resources available, and the nature of long-term wildlife, timber, and aesthetic goals. The number of oak stems established per acre and their growth rates may take a back seat to the number of oak species established if wildlife goals have the highest priority, while the reverse would be true if oak timber production is the main concern. Artificial oak regeneration in areas of the country with very high deer populations is guaranteed to be more expensive than in areas with low deer populations, and greater expenditures on artificial oak regeneration are likely as the mature oaks in current oak-dominated forests continue to age past their reproductive prime and succumb to various sources of mortality.

Further suggested reading

Buckley, D.S. 2002. Field performance of high-quality and standard northern red oak seedlings in Tennessee. In Outcalt, K.W., ed. *Proceedings of the Eleventh Biennial Southern Silvicultural Research Conference*, Knoxville, TN, 2002, pp. 323–327. USDA Forest Service Southern Research Station General Technical Report SRS-48, Asheville, NC, 622 pp.

Buckley, D.S. and T.L. Sharik. 2002. Effect of overstory and understory vegetation treatments on removal of planted northern red oak acorns by rodents. *Northern Journal of Applied Forestry* 19:88–92.

Buckley, D.S., T.L. Sharik, and J.G. Isebrands. 1998. Regeneration of northern red oak: Positive and negative effects of competitor removal. *Ecology* 79:65–78.

Burger, J., D. Graves, P. Angel, V. Davis, and C. Zipper. 2005. The forestry reclamation approach. Appalachian Regional Reforestation Initiative Forest Reclamation Advisory No. 2, 4 pp. arri.osmre.gov/FRA/Advisories/FRA_No.2.7-18-07. Revised.pdf.

Burns, R.M. and B.H. Honkala (tech. coords.). 1990. *Silvics of North America, Vol. 2: Hardwoods, Agriculture Handbook 654.* U.S. Department of Agriculture, Forest Service, Washington, DC, Vol. 2, 877 pp.

Clark, S.L., S.E. Schlarbaum, and P.P. Kormanik. 2000. Visual grading and quality of 1-0 northern red oak seedlings. *Southern Journal of Applied Forestry* 24:93–97.

Groninger, J.W. 2005. Increasing the impact of bottomland hardwood afforestation. *Journal of Forestry* 103:184–188.

Hodges, J.D. and G.L. Switzer. 1979. Some aspects of the ecology of southern bottomland hardwoods. In *North America's Forests: Gateway to Opportunity*, 1978 Joint Convention of the Society of American Foresters and the Canadian Institute of Forestry, St. Louis, MO, October 22–26, 1978, pp. 360–365. Society of American Foresters, Washington, D.C.

Johnson, P.S. 1979. Shoot elongation of black oak and white oak sprouts. *Canadian Journal of Forestry Research* 9:489–494.

Johnson, P.S., S.R. Shifley, and R. Rogers. 2009. *The Ecology and Silviculture of Oaks*, 2nd edn. CABI, New York.

Kormanik, P.P., S.S. Sung, and T.L. Kormanik. 1994. Toward a single nursery protocol for oak seedlings. In *Proceedings of the 22nd Forest Tree Improvement Conference*, Atlanta, GA, June 14–17, 1993, pp. 14–17.

Kriebel, H.B. 1976. *Twenty-Year Survival and Growth of Sugar Maple in Ohio Seed Source Tests.* Ohio Agriculture Research and Development Center, Research Circular 206, University of Ohio, Wooster, OH.

Lockhart, B.R., B. Keeland, J. McCoy, and T.J. Dean. 2003. Comparing regeneration techniques for afforesting previously farmed bottomland hardwood sites in the Lower Mississippi Alluvial Valley, USA. *Forestry* 76:169–180.

Oswalt, C.M., W.K. Clatterbuck, and A.E. Houston. 2006. Impacts of deer herbivory and visual grading on the early performance of high-quality oak planting stock in Tennessee, USA. *Forest Ecology and Management* 229:128–135.

Reich, P.B., R.O. Teskey, P.S. Johnson, and T.M. Hinckley. 1980. Periodic root and shoot growth in oak. *Forest Science* 26:590–598.

Stanturf, J.A., E.S. Gardiner, J.P. Shepard, C.J. Schweitzer, C.J. Portwood, and L.C. Dorris Jr. 2009. Restoration of bottomland hardwood forests across a treatment intensity gradient. *Forest Ecology and Management* 257:1803–1814.

Teclaw, R.M. and J.G. Isebrands. 1986. Collection procedures affect germination of northern red oak (*Quercus rubra* L.) acorns. *Tree Planters' Notes* 37:8–12.

chapter nine

Fire in the oak woods
Good or bad?

Craig A. Harper and Patrick D. Keyser

Contents

How did this happen: Doesn't fire kill hardwoods?............................... 101
So, why was fire suppressed, and why are we so hesitant to use fire?..... 101
The role of fire in eastern oak systems... 104
Prescribed fire versus wildfire: What is the difference as related to
strategies for managing oak systems?.. 105
Fire effects on vegetation and wildlife habitat............................... 105
Fire effects on forest soils and water sources................................ 107
Fire effects on wildlife... 108
Using fire intensity, timing, and frequency to meet management
objectives... 109
Weather and fuel considerations...111
Atmospheric stability 113
Considerations when using fire in oak systems 113
   Community restoration 114
   Oak regeneration and wildfire prevention...................................114
   Wildlife considerations...114
      White-tailed deer...115
      Black bear...115
      Other mammals 116
      Wild turkey..116
      Ruffed grouse 117
      Northern bobwhite..117
      Forest songbirds...118
      Old-field and shrubland songbirds....................................119
      Reptiles and amphibians 119
      Livestock..120
Implementing prescribed fire in oak systems......................................121
   Preparing the site..122
   The burn plan...122
   Notifying appropriate contacts prior to burning.............................123

Firing techniques... 123
    Backing fire.. 124
    Strip-heading fire... 124
    Flanking fire 125
    Point-source fires... 125
    Ringfires or center fires.. 125
Smoke management considerations.. 126
Post-burn evaluation 127
Human perception: Opportunities and limitations 128
Further suggested reading.. 129

The historical occurrence and role of fire in oak-dominated forests, woodlands, and savannahs are well documented. Fire, whether spread by humans or via lightning, directly influenced both the vegetation composition and the structure of these systems for thousands of years prior to widespread fire exclusion in the nineteenth and twentieth centuries. Research throughout the Eastern United States has shown fire within upland oak-dominated systems commonly occurred every 25 years, and within every 4–6 years on many sites. Fire occurred during all seasons. Some studies indicate fire was most prevalent in these systems during the dormant season, whereas others show a peak during May or during the fall after leaf senescence. The vast majority of lightning-caused fires occur during April–August in the Eastern United States, whereas fires spread by people, either Native Americans or European Americans, have been most prevalent in the spring and fall (Figure 9.1).

Figure 9.1 Oak forests, woodlands, and savannahs in the Eastern United States are well adapted to fire. (Photo courtesy of C. Harper.)

How did this happen: Doesn't fire kill hardwoods?

Pollen records in the Eastern United States show areas now dominated by oak forests and woodlands have contained oaks for thousands of years. What has been lost more recently, however, are many of the herbaceous species prevalent before widespread fire exclusion. Without question, upland oak species that occur in this region are adapted to fire. Yes, fire can kill oaks. However, fire can also kill pines or any other tree species. Thick bark resists the heat of fire and protects cambium tissue (the thin inner-bark layer where nutrients are transported from the roots to the limbs). When the bark splits and cambium tissue is damaged, a wound develops that enables insects and disease to enter the tree. Young trees (of all species) are most susceptible. Older trees with thicker bark are more resistant. Some southern yellow pines (such as longleaf, shortleaf, loblolly) produce very thick bark, which accumulates in layers as the tree ages, protecting them from heat. Most deciduous trees have relatively thin bark and are not as resistant to fire. Upland oaks, however, have relatively thick bark and are more resistant to heat than other hardwood species, such as yellow poplar, sweetgum, and maples. Upland oaks also resist wounding and decay as they are able to heal more quickly or completely than many other deciduous trees. One study showed wounds in oaks following fire were closed within 1–2 years and produced only limited discoloration of the wood. Oaks also have an advantage of developing relatively deep root systems as seedlings. When the top is killed by fire, the seedling root systems contain enough energy to re-sprout and grow a considerable stem the following growing season. Nonetheless, even with these fire-adapted strategies, oaks are more susceptible to fire than most southern yellow pine species, and fire-damaged oaks have less timber value, which is a real consideration for many landowners (Figure 9.2).

So, why was fire suppressed, and why are we so hesitant to use fire?

Fire was used by Native Americans throughout the Eastern United States for thousands of years. This practice was continued, especially in the American South, by European settlers of Scottish and Irish descent who largely made their living by free-ranging livestock. Woods burning was conducted annually throughout the South to improve grazing conditions in forests and woodlands. It should be noted that burning was widespread and important for the economy. Just prior to the War Between the States, livestock sales in the South equaled those of cotton and all other crops combined. Fencing largely was illegal at this time ("free-range" laws) and burning often was viewed as a community affair to improve common range. Using fire was not learned by settlers from native Americans but

(a)

(b)

Figure 9.2 Fire can damage oaks as well as other tree species. Damage to oaks often is associated with intense wildfire or when large debris burns at the base of the trunk. Debris burning retains heat longer and increases the likelihood of damaging the cambium layer and creating a wound susceptible to insects and disease. Low-intensity fires do not harm overstory trees, especially if large debris is moved a few feet from the base of the tree. (Photo courtesy of C. Harper.)

was a continuation of traditions practiced by Celtic peoples. Controlled burning was not practiced as much in the Northeast or Upper Midwest because settlers in those regions primarily were from the southeastern lowlands of England and continental Europe where forests were dominated by fire-sensitive hardwoods and spruce. Common agricultural practices in those regions were row cropping on small acreage and fenced pasture with a few livestock.

As the forestry profession began to emerge in the late nineteenth and early twentieth centuries, conflicts developed between foresters and those who ranged livestock. Foresters educated and trained in Europe and in the Northern United States did not understand and disdained the practice of woods burning. While working in North Carolina, Gifford Pinchot and W.W. Ashe (1897) stated "The first and absolute prerequisite before any attempt can be made to improve the condition of the longleaf pine forests is entire exclusion of cattle and hogs and complete protection from fire." Obviously, they did not understand the longleaf pine ecosystem. These sentiments also led to the exclusion of fire from all other forest types as well, including those dominated by oaks.

During the 1920s and 1930s, an all-out campaign was waged by the federal government to curtail the use of fire across the landscape. State forestry agencies received funding through the Clarke–McNary Act of 1924, but these funds were withheld from states if they tolerated controlled burning. In 1927, the American Forestry Association employed the "Dixie Crusaders" through the Southern Forestry Education Project to "educate" rural Southerners in an effort to end the practice of woods burning. Millions of brochures and posters were handed out and thousands of movies were shown by these misinformed propagandists. Interestingly, these movies showed scenes from disastrous crown fires in the Western U.S. and Lakes States, not from the South where controlled burning had been practiced successfully for generations. The U.S. Forest Service even hired sociologists and psychologists to study why Southerners used fire. This information was then used in one of the most successful advertising campaigns in history: Smokey the Bear. The use of scriptures from the Bible and the images of Smokey the Bear praying for people to not use fire were eventually successful in making the American public believe fire in the woods was always bad.

During this time, there were some scientists and professional foresters and wildlife biologists who recognized the importance of fire in maintaining and restoring ecological communities. Their advocacy of the use of fire, however, was fiercely opposed and discredited by those in power within forestry agencies. Over time, however, more professionals began to acknowledge the need for fire, particularly in pine systems of the Coastal Plain. Herbert Stoddard championed this movement as he chronicled the need for fire to maintain and restore the longleaf system and the importance of fire when managing for northern bobwhite. By the 1950s, the U.S. Forest Service cautiously began to use "prescribed fire" to reduce fuel loads and enhance wildlife habitat. [It is interesting to note that as this was happening, forestry agencies insisted that only "experts" must "prescribe" the practice and that "controlled burning" still was not recommended!] During the 1960s and 1970s, prescribed fire was being used more, and even a little outside the Coastal Plain. It also was during this time that the National Park Service reversed their stringent anti-fire policy to allow burning where necessary to restore vegetation communities that had been degraded or almost totally lost during four decades of fire suppression.

By the 1990s, the use of fire in upland hardwoods had been recognized both for forestry and wildlife applications. Indeed, the use of fire in hardwood systems is different from that in pine systems. Still, there is much to learn about the effects of fire on various hardwood systems. However, considerable research has shown fire has a place in oak-dominated systems and must be used to restore some community types, such as oak savannahs and woodlands.

The role of fire in eastern oak systems

As in other systems, fire plays an important role in reducing fuels, recycling nutrients, and influencing plant community composition and structure in oak systems. As the plant community changes, so does the associated wildlife community. Fire is used in oak systems not only for ecological restoration but also for specific forest and wildlife management objectives.

Fire can be used to influence the amount of sunlight that enters a stand prior to a regeneration harvest and stimulate advance oak regeneration. Even relatively low-intensity fire can triple the amount of light (often from approximately 5%–15% sunlight) entering a closed-canopy stand as woody stems in the understory and some in the midstory are top-killed. After a regeneration harvest, fire may be used to help control competition, such as yellow poplar and maple sprouts, and allow young oak root systems to send up a vigorous sprout that may better compete with future competition arising from seed.

Objectives in wildlife management as related to the use of fire often include stimulating the seed bank to encourage fresh understory growth, increase herbaceous forage, increase soft mast production from understory forbs, brambles, and shrubs, and influence the structure of the understory to meet the requirements of various wildlife species. Fire intensity, frequency, and timing all can be adjusted to meet various objectives.

Fire exclusion leads to an unnatural plant composition and structure. It has been said, accurately, that the woods are going to burn; it is just a matter of when and how hot. When fire is excluded, many plant species disappear and others dominate that would not otherwise. As fuel loads build, the stage is set for a relatively extreme wildfire. Unfortunately, this is common in the Western United States. Across the country, prescribed fire is now being used proactively to restore several ecological communities that have declined to the point they are nearly gone. In the Eastern United States, this is particularly true of oak savannahs and woodlands. The remainder of this chapter will discuss the effects of fire on the oak systems in the Eastern United States and how fire can be used in these systems to achieve various management objectives.

Prescribed fire versus wildfire: What is the difference as related to strategies for managing oak systems?

Prescribed fire and wildfire can have the same characteristics, or they can be very different. Prescribed fire is *fire applied in a knowledgeable manner on a specific area under selected environmental conditions to accomplish predetermined management objectives*. Wildfire may occur under these same conditions, but it also may occur under more extreme conditions and produce undesirable results. Prescribed fire is not always of low intensity. There are times when high-intensity fire is needed to achieve management objectives. However, even then, prescribed fire is applied under selected environmental conditions to accomplish a predetermined management objective.

Wildfire does not always produce negative effects. In fact, in many (if not most) cases, the environmental effect of wildfire is positive. Most wildfires, especially in the Eastern United States, are not raging fires out of control. Slow-moving flames, less than 1 foot tall, are more common. In many cases, this is purely natural and is producing positive effects for the forest community. However, over the past several decades, society has been "programmed" to think all fires in the woods are bad, and the process of changing this way of thinking will be slow, probably generational. To their credit, many agencies now allow wildfires that pose no problem to burn where there is no possible damage to homes and other structures and where there are no problems with the associated smoke.

Fire effects on vegetation and wildlife habitat

Fire can be used in oak systems to influence vegetation composition and structure, which can influence regeneration as well as encourage additional food resources and enhance cover conditions for a wide variety of wildlife species. Fire sets back existing vegetation and stimulates fresh growth from the seed bank and from sprouting. This disturbance may influence growth/competition of different tree species, and the new vegetative growth is more palatable and nutritious for grazers and browsers as plants are more digestible and contain higher nutrient levels following burning, especially early in the growing season following fire. Fire consumes the litter layer, which not only stimulates the seed bank but also makes food resources, such as seed, mast, and invertebrate parts, in the leaf litter more readily available to wildlife.

Fire typically leads to increased herbaceous forage and browse, as well as soft mast, but this is directly related to the amount of sunlight penetrating the canopy. In closed-canopy stands, the increase in forage and browse may be negligible unless additional sunlight reaches the forest floor as a result of opening the canopy. Allowing at least 20%–30% sunlight into the stand through an improvement cut or some type of thinning prior to or soon after burning is necessary if increased forage/browse is desired. Increased light availability can also improve soft mast production. Burning helps scarify seed and encourage germination of some plants that produce soft mast, but if sufficient sunlight is not available, an increase in soft mast may not be realized. Blackberry, pokeweed, and blueberry are often the primary soft mast species responding after canopy reduction and fire in oak forests in the Eastern United States. Pokeweed usually responds the growing season following disturbance, but blackberry and blueberry require at least 2 years before an increase in soft mast production is realized. Soft mast production often begins to decline after 5 years, and by 7 years post-disturbance, the regenerating woody stems in the understory usually have developed to the point they are shading out soft mast species, and additional prescribed fire is needed to control competition from woody saplings and stimulate additional soft mast production. Invertebrate abundance is tied to the leaf litter layer and understory structure and composition. It is not unusual for invertebrate abundance to decline after fire. However, the availability of invertebrates for ground-feeding birds may be increased if there is increased groundcover that provides ground-feeding birds overhead protection from predators while foraging. Thus, the *availability* (not necessarily *abundance*) of invertebrates, as well as other foods, can be enhanced following fire (Figure 9.3).

Food availability is important for wildlife, but arguably more important are the structure and the cover present. Various wildlife species require different amounts of groundcover, vertical cover (layers of vegetation from the ground up), and horizontal cover through various stem densities. Until the structure is suitable for a particular wildlife species, food availability may be meaningless. Fire can be used to create the desired structure within a stand for various wildlife species. However, just as related to food, light availability is critical to influence the structure within the stand. With sufficient sunlight, groundcover is stimulated by fire. This is desirable for species that forage at ground level, for those that

Figure 9.3 The canopy in this stand was reduced to allow approximately 30% sunlight to reach the forest floor. It has been burned three times on a 2- to 3-year fire-return interval. The structure of the understory has improved dramatically, providing increased food and cover for many wildlife species. (Photo courtesy of C. Harper.)

nest in understory vegetation, and those that require groundcover for raising young. Without additional fire, the understory woody response soon grows large enough to become a part of the midstory. As mentioned earlier, within about 7–8 years (in the Eastern United States where rainfall is not limiting), the amount of sunlight penetrating the midstory is reduced to that available in a closed-canopy situation (usually 0%–5% sunlight). The structure at this point may be good for some species, such as gray catbirds, white-eyed vireos, and ruffed grouse, but not for others, such as summer tanagers, great-crested flycatchers, and wild turkeys. At this point, your management strategy should follow your objectives. If you are trying to restore an oak savannah or woodland, you certainly need to continue to burn. If you have considerable acreage and you are concerned with regenerating the stand, it may be desirable to allow the regeneration to develop. But if your acreage is limited, and your objective is wildlife that would benefit from increased groundcover, soft mast, and visibility, you should consider additional fire to foster the desired conditions.

Fire effects on forest soils and water sources

Fire intensity is key to how soil and water resources are affected by fire. Low-intensity fires have no negative effect on forest soils or water resources. Organic matter in the upper soil layer is generally unchanged, but slight increases have been reported. As plant material is consumed, nutrients are released as rain leaches them back into the soil where they are available for new plant growth. Although nitrogen is lost to the atmosphere when fuels are consumed, nitrogen actually may be increased on the site following fire through nitrogen fixation. Intense fire, however, reduces soil organic material and may reduce soil fertility. Burning when soil and/or fuel moisture conditions are extremely low can lead to reduced soil fertility and also damaged root systems. This is common when burning debris piles. When the duff layer of the forest floor is consumed and mineral soil is exposed, soil infiltration and aeration may be decreased as soil pores are clogged with fine particles of soil and carbon. When this condition exists, surface runoff and soil erosion are more likely. In most circumstances, burning should be conducted when the duff layer under the leaf litter is relatively moist. When the duff layer remains intact, the concern of surface runoff and sedimentation into adjacent water sources is removed.

Fire effects on wildlife

Effects of fire on wildlife are usually indirect. That is, wildlife is affected by changes in vegetation, food availability, and habitat composition and structure. Changes in habitat quality and arrangement then influence daily movements and home range sizes. These changes, thus, can influence reproductive rates, recruitment, and survival. Rarely are wildlife killed when prescribed fire is conducted in a sensible manner. Intense fire, whether wildfire or prescribed, presents more danger for wildlife. Not only does increased fire intensity pose more danger, but the firing technique used, as discussed in the following, must be considered where wildlife is a concern.

Relatively low-intensity prescribed fire rarely poses danger to wildlife. Large and medium-sized mammals, such as white-tailed deer, black bear, gray or red fox, and raccoon, leave the site. Small mammals, such as chipmunks or squirrels, hide underground or in trees. Birds fly. Reptiles and amphibians often escape underground or under large debris and objects, such as logs and rocks. However, mortality of some amphibians and reptiles can be considerable, especially if burning is conducted soon after their emergence from winter hibernacula when they are lethargic. Disturbance of nests and young often is a concern when implementing prescribed fire. As discussed in the following, these concerns can be alleviated by adjusting the timing of burning to occur outside the nesting or young-rearing seasons and not when slow-moving animals are emerging from hibernacula.

Not all species benefit from fire. Some species, such as ovenbird, require leaf litter for nesting material and a substrate for feeding.

Many amphibians, particularly the lungless salamanders, require a relatively moist environment for respiration as they actually breathe through their skin. If additional sunlight reaches the forest floor as a result of midstory or overstory reduction caused by burning, their environment can become excessively dry. This, however, rarely will be a concern based on use of fire alone—unless it is used repetitively over a number of years and a more open woodland condition is created. Fire, just as other disturbance events, should be implemented on a scale that is reasonable and acceptable for a variety of wildlife species. When a stand is burned or harvested, wildlife species that are negatively affected are able to use adjacent stands or other areas that were not burned or harvested. They are not necessarily "destroyed" or killed just because the environment they have been living in has been altered. What is good for one species is never good for all species. Implementing an appropriate balance of successional stages and vegetation types through fire and other disturbance practices across the landscape to benefit a diversity of wildlife species should be considered.

Using fire intensity, timing, and frequency to meet management objectives

Fire intensity, timing, and frequency strongly influence fire effects. Fire intensity is related to fuel loads and environmental conditions. Hotter, more intense fires kill larger trees than cooler, low-intensity fires, which only may kill small woody stems in the understory and some small-diameter (<3 inches at ground level) stems in the midstory. The cambium layer of shrubs and trees is severely damaged when heated to approximately 145°F, and this normally kills the aboveground stem. For larger trees with thicker bark, a more intense fire is needed to heat the cambium layer to a level that causes damage. Increased heat can also result if debris (dead limbs, slash) is lying next to or lodged against the trunk. As the debris burns, the heat is retained around the trunk longer than if only the surrounding leaf litter was burning. This extended heating period can cause damage to the cambium on the side(s) where the debris is burning. This usually results in scarring ("cat faces") and presents opportunity for insects and diseases to enter the tree. This also results in wood discoloration and degrades timber value.

Low- to moderate-intensity fire is prescribed for most applications in oak-dominated systems. Only when a significant reduction in overstory cover is desired and there is no consideration for timber value, should intense fire be recommended in oak systems. Even when creating oak woodlands or savannahs (see Chapter 15), it is not recommended to use intense fire because desirable residual trees are likely to be damaged. A better approach would be to mechanically remove or kill undesirable trees via girdling and herbicides, and then use low- to moderate-intensity fire at the correct time to help reduce woody competition and encourage more herbaceous groundcover.

Fire intensity always is a concern in stands that have never been burned and especially in those that have been logged within the past 2–3 years. Litter accumulation may be deep, and there is usually a considerable amount of debris in the stand. That is, there is a lot of accumulated fuel ready to burn and conditions are set for a relatively intense fire. When burning a stand for the first time, it is critical to make sure there is adequate moisture in the leaf litter. Burning techniques to allow a low-intensity fire are discussed below. The prudent fire practitioner burns only under the correct environmental conditions that allow fire intensity to be controlled and to meet landowner objectives.

Short fire-return intervals (1–2 years) generally increase structure near the ground (<4 feet) and reduce the amount of structure 4–12 feet tall. Fire-return intervals of 3–5 years maintain a diverse vegetation structure on most sites from the ground to approximately 12–15 feet. However, structure is highly dependent on the amount of light entering the stand. More light will allow more dense structure. Long fire-return intervals (>5 years) in closed-canopy stands may have no impact on vegetation composition or structure, unless fire intensity is great, which would kill more midstory stems and could kill some overstory trees, breaking the closed canopy and allowing increased light to the forest floor. If dormant-season fire is used, an understory dominated by woody re-sprouts may persist. Carbohydrates are stored in the root systems of woody plants during winter senescence. Dormant-season fire may top-kill the plant, but the root system is not harmed. Upon spring green-up, the plant is able to use stored root reserves to produce new sprouts, which may eventually grow into the midstory if not subsequently disturbed. Early growing-season (April/May) fire may be more effective than dormant-season fire but does not control woody species as effectively as those in the late growing season. As the growing season progresses and a majority of carbohydrates are transported from the root system to the aboveground portion of the plant, the shrub or tree is more susceptible to fire. If fire top-kills the plant during this time, it is more likely to die and not re-sprout. In stands where the objective is to reduce woody and increased herbaceous cover in the understory, a late growing-season (September/October) fire will generally produce more favorable results (Figure 9.4).

Relatively infrequent fire (beyond every 6–8 years) will allow woody species to remain dominant in the forest understory. However, according to the timing of fire, fire frequency can be adjusted for various objectives. When using late growing-season fire, it may not be necessary to burn as often to maintain a significant herbaceous component in the understory than if dormant- or early growing-season fire is used. An exception to this

Figure 9.4 Late growing-season fire may be used to reduce woody composition and promote forbs and grasses in the understory. Dormant-season fire is less effective in achieving these same results. (Photo courtesy of C. Harper.)

is when managing oak savannahs. Once suitable vegetation composition in a savannah has been developed, dormant-season fire can be used to maintain this composition longer than if managing a forest stand where woody species dominate. When perennial native grasses and forbs are well established in a savannah, they can be perpetuated with dormant-season fire. Growing-season fire still is recommended occasionally, however, to help maintain vegetation diversity. A common recommendation may be to burn the savannah on a 2–4-year fire-return interval. Two or three dormant-season fires may be used for every late growing-season fire. Strict burning regimes should be avoided. It is best to be flexible and use fire in response to vegetation changes and objectives.

Weather and fuel considerations

Wind speed, relative humidity, temperature, rainfall, and atmospheric stability are the primary factors that influence fuel moisture and fire intensity. They also affect decisions to burn as related to smoke management concerns. As you learn to burn, you soon will realize it is absolutely critical to follow weather patterns and forecasts when planning to burn. Accurate weather forecasts for planning prescribed fire can be obtained at the National Weather Service and state forestry agency websites. The importance of monitoring the weather cannot be overstated. It should be a daily routine when you are planning or needing to burn. Weather patterns and forecasts should continue to be monitored in the days leading to burning and right up until burning is initiated. Local (on-site) conditions can be monitored by hand-held instruments that provide temperature, wind speed, and relative humidity readings. These instruments are available through forestry supply dealers. Burning must be given top priority among management activities when burning weather approaches. The correct weather and fuel conditions to enable your fire to meet your objectives are usually short lived, and you must be ready when those conditions are present.

Predictable, steady winds are desirable when burning. Winds are most predictable during the dormant season after passage of a cold front when persistent winds, low relative humidity, and cool temperatures during sunny days are all but assured. Weather conditions during the growing season are less predictable and reliable. Predicted wind speeds are often in excess of on-the-ground winds. It is important to note wind speeds are forecasted for open areas 20 feet aboveground. Actual in-stand (inside the woods) wind speeds 6 feet aboveground are considerably lower. In-stand wind speeds at eye level should be 1–5 mph for safe, predictable burning conditions. Wind speed and direction are influenced by many things in addition to the current weather pattern. The topography and arrangement of roads and openings in the stand can influence wind speeds and direction and should be considered prior to and when burning.

Relative humidity is the amount of moisture in the air compared to the total amount of moisture the air is capable of holding at that temperature. Relative humidity is an extremely important parameter when burning. It readily influences the flammability of fine dead fuels, which respond quickly to changes in the relative humidity. The preferred relative humidity for a majority of burning prescriptions is 30%–40%. Relative humidity levels below 30% can be dangerous as fires are more intense and unintended spot fires are more likely. Burning when the relative humidity is above 40% is often more patchy and may not burn complete enough to reach desired objectives. However, depending on the management objectives (unburned patches creating a mosaic pattern) and fuel load and composition (lots of fine fuel, especially grasses), successful burning may be conducted when relative humidity is above 50%.

Ambient air temperature is an important factor when burning since higher temperatures dry fine fuels more quickly than cooler temperatures, and higher temperatures increase the likelihood a fire will reach lethal temperatures to kill vegetation (145°F at the cambium layer). This has real implications when burning during the growing season. That is, a relatively cool, low-intensity fire during the growing season is more likely to reach lethal temperatures to kill woody saplings than a relatively cool, low-intensity fire during the dormant season.

Like wind, rainfall is more easily predicted in the dormant season as opposed to the growing season when thunderstorms and showers can arrive unexpectedly. The main consideration with rainfall is soil and fuel moisture. It is important to burn only when there is adequate soil and fuel moisture to protect against soil erosion and sedimentation and ensure root systems are not damaged (unless your objective is to kill some overstory trees). As mentioned earlier, when the duff layer under the leaf litter is left intact, the danger of erosion, sedimentation, and soil impoverishment is minimized. It often is possible to burn fairly soon after rain. Openings and savannahs often can be burned the day after a rain, especially if there is considerable coverage of dead grass litter, and it is not unusual for conditions to permit burning in forest stands within 3 days after a rain, particularly during the dormant season when considerable sunlight is reaching the forest floor and air flow at ground level is more likely.

Atmospheric stability

Atmospheric stability is the resistance of the atmosphere to vertical motion. When burning, it is desirable for smoke to rise and disperse rapidly. When the atmosphere is very stable, temperatures decrease slowly as altitude increases, and the mixing height (where vigorous mixing of the atmosphere occurs) is relatively low. Burning should not be conducted under those conditions. A desired mixing height is at 1500 feet aboveground, where the temperature is increasing more than 5.5°F per every 1000 feet in altitude. A neutral or slightly unstable atmosphere with a relatively high mixing height and adequate transport wind speeds (more than 9 mph) is desirable to effectively move smoke away from the site, not restrict visibility, and not impact air quality. Forecasts for mixing heights and transport wind speeds are given as part of fire weather forecasts provided by the National Weather Service and state forestry agencies.

Considerations when using fire in oak systems

There are many issues to consider prior to using fire in oak systems, but the first is to clearly define your goals and objectives and prepare a written management plan with a professional forester and wildlife biologist (see Chapter 13). Is regeneration and/or forest health the primary consideration? Have fuels accumulated to dangerous levels? Are you trying to establish an oak savannah or woodland? Is wildlife a primary consideration? If so, which species? There is nothing more ambiguous than to say you want to manage for "wildlife!" All species have different habitat requirements, therefore management efforts for one species may be completely counter to another. Your goals, objectives to reach your goals, the strategy and methods you intend to employ to reach your objectives, and a timeline for your efforts should be stated clearly in a management plan. This is very important and an often overlooked step that can lead to confusion, frustration, and incomplete implementation.

Discussed in the following are some factors to consider when using fire for various objectives. Several of these issues are discussed in more detail in other chapters of this book. The intention here is to compare/contrast how fire might be used differently, depending on management objectives.

Community restoration

Using fire is essential if you are trying to restore or establish an oak woodland or savannah (see Chapter 15). Continued disturbance using fire is required to set back woody stem density and establish or maintain an understory plant community dominated by herbaceous species. Fire and how it is used are the drivers of these systems. A diversity of plant species and structure is achieved and maintained by using fire at different times of the year and with varying levels of intensity. Additional considerations regarding mechanical removal, basal area, and herbicide applications when managing oak savannahs and woodlands are discussed in Chapter 15.

Oak regeneration and wildfire prevention

Where oak regeneration is the primary objective within a forest stand, fire can be used to reduce understory and midstory stem density of competitors and slightly increase available light to stimulate advance oak regeneration. Once advance oak regeneration is present, various regeneration methods (such as shelterwood) may be employed (see Chapter 7). Growing-season fire also has been used to control fast-growing competition (such as yellow poplar and maples) 2–4 years following an initial shelterwood harvest. Setting the competition back and allowing the strong root reserves of oak seedlings to produce a tall main stem the following growing season can help oak regeneration get a competitive advantage and become established in the regenerating stand.

Occasional fire in forest stands can also prevent dangerous fuel loadings. Using low-intensity prescribed fire every 4–5 years in closed-canopy forests will reduce fuel loads, rejuvenate understory production, and help prevent wildfire that could damage desirable overstory trees.

Wildlife considerations

Fire can be used to enhance conditions for many wildlife species. However, because all species have different habitat requirements and occupy a unique niche in the system, fire is not used in the same way to benefit all species. It is beyond the scope of this chapter to provide complete descriptions of habitat requirements for all species, but the primary effect of fire and its use for several species and species groups are provided below. Additional information and considerations for various wildlife species can be provided and summarized by a Certified Wildlife Biologist® in your management plan.

White-tailed deer

Fire can be used in oak systems to increase both available nutrition and cover for white-tailed deer. Burning in closed-canopy forest will do little to improve browse availability or fawning cover. Therefore, it is important to reduce canopy closure and allow at least 20%–30% sunlight into the stand prior to or soon after burning. Following an improvement cut, thinning, or shelterwood harvest, forage selected by white-tailed deer can increase from approximately 25–100 to 700–1000 lb (dry weight) per acre. Not only can this help increase the nutritional carrying capacity of the area for deer, but it also provides additional cover for fawning. The soft mast response following fire also provides increased nutritional benefits for deer. Interestingly, the species that provide the majority of soft mast in oak systems of the Eastern United States (blackberry, blueberry, pokeweed) are also important forage species and can provide excellent fawning cover.

Dormant-season fire and growing-season fire can be used to stimulate increased forage, soft mast, and cover for deer. Dormant-season fire with a fire-return interval of 3–5 years in oak forests will maintain browse availability before declining. Past 5–7 years, browse will become less available as it grows beyond the reach of deer and shading reduces groundcover. In woodlands and savannahs, a 2- to 4-year fire-return interval will maintain browse availability. Late growing-season fire may encourage more forbs where grasses dominate (>50% cover). Increased forb cover (at least 30% groundcover) is desirable when managing for deer because forbs are preferred forage for deer during the growing season. Woody and/or forb composition may comprise 30%–70% of the groundcover. Grasses are not desirable for deer forage, but some grass cover may be needed to burn the area without excessive wind. It is important to distribute disturbance spatially across a property and landscape when managing for white-tailed deer.

Black bear

Fire can be used to increase available foods (especially soft mast) for black bear as well as dense cover. In general, a longer fire-return interval may be desirable when managing for black bear because of the need for increased soft mast and dense cover. The soft mast response following fire may persist 7–8 years post burning. Dense cover is also available during that period, depending on the amount of sunlight present. Movements and home range sizes of black bears correspond directly with food availability. When increased food is available, movements and home range sizes are decreased substantially, which leads to increased production in the bear population. Using fire to manage oak forests that allow sufficient sunlight to the forest floor to provide increased soft mast and dense cover via regenerating stems, and providing this across a landscape, is an important strategy when managing for black bears.

Other mammals

As fire influences the vegetation community, food and cover resources are impacted for many species. With increased groundcover productivity in oak forests, woodlands, and savannahs, there typically is a response in the small mammal community. Although perhaps no one manages property specifically for small mammals, the abundance and distribution of rodents are critical for many other species. Not only do small mammals provide food (prey) for predators (such as bobcat, coyote, gray fox, and raccoon), they also affect other prey species (such as wild turkey, ruffed grouse, and northern bobwhite). Predators not only benefit from a diverse small mammal population, many also benefit from the soft mast response following fire, such as coyote, gray fox, skunks, opossum, and raccoon. The vegetative cover influenced by fire also has real implications for other mammals, especially rabbits, groundhog, and bats using oak savannahs.

According to the species of interest, fire may be used in different seasons and with varying return intervals to meet management objectives. However, in general, when influencing predator–prey relationships across multiple species and different successional stages, using fire in different seasons in different intervals and with varying intensity is recommended.

Wild turkey

Wild turkeys benefit from relatively low woody cover for nesting and more open herbaceous cover for raising broods. Wild turkeys desire relatively open cover for enhanced visibility, especially above 2 feet. Prescribed fire can be used to facilitate these needs. Both dormant- and growing-season fire in oak forests, woodlands, and savannahs can be used to establish and maintain desirable cover requirements and provide additional food resources.

Closed-canopy forests may be burned to improve foraging conditions for wild turkeys, but this may not improve cover appreciably for nesting or brooding. Reducing canopy cover through some type of thinning or improvement cutting prior to or soon after burning usually is necessary to improve cover for nesting and/or brooding. Where sufficient sunlight is available, dormant-season fire on a 3- to 5-year interval will maintain desirable structure for nesting and brood rearing and provide soft mast. Where desirable, especially in woodlands and savannahs, occasional late-growing season fire should be used to promote additional forb cover and decrease woody composition where needed.

Using fire during the early growing season overlaps with the nesting season, which has caused concern among some wildlife managers. This concern can be alleviated by burning during the dormant season, which influences vegetation composition similar to early growing-season fire. Nonetheless, it can be difficult to complete all burning necessary on large properties in some years because of insufficient human resources or incompatible weather patterns. Thus, it is often necessary (and desirable) to conduct some burning during the growing season. Although some nests may be destroyed by early growing-season fire, the abundance and success of nests in the areas that have been burned in the past few years (where cover is better) may compensate for any losses. Wild turkeys also readily re-nest when their nests are destroyed during the egg-laying period or early in the incubation period. All that being said, where nest destruction is a concern for wild turkeys and you have not completed all burning necessary during the dormant season, late growing-season fire may be used instead of early growing-season fire, and usually with better results with regard to reducing woody stem density and encouraging more herbaceous cover.

Ruffed grouse

Ruffed grouse are forest birds that require young stands (5–20 years old) with relatively high stem densities. They also may use mature stands with lush herbaceous cover for brood rearing during the early summer and to find food (especially acorns) during the fall/winter.

Dormant-season prescribed fire can be applied every 6–8 years in mature stands that provide adequate sunlight to enhance understory cover and soft mast production. Unlike wild turkeys, ruffed grouse do not readily re-nest, especially in the Central and Southern Appalachians where oak forests predominate. Therefore, burning during the early growing season should be avoided where ruffed grouse is a focal species. Where ruffed grouse is the primary objective, burning young stands to stimulate sprouting and maintain high stem densities every 10–15 years is desirable. This burning regime is especially effective following a improvement cut or two-aged shelterwood that leaves approximately 20–40 square feet of basal area comprising both red and white oak species of mast-bearing age. These stands provide suitable cover and structure, as well as a high-quality food resource during the winter, all in the same stand.

Northern bobwhite

The northern bobwhite *depend* on disturbance. In fact, the bobwhite quail has been called the "fire bird" by many wildlife managers and scientists. Bobwhite require relatively open landscapes, optimally dominated by native forbs and grasses with shrub cover well interspersed throughout. The northern bobwhite is not a woods bird, and if found frequenting closed-canopy woods with any regularity, it can be accurately assumed that conditions for bobwhite on that property are poor and the quail population is in a steady decline. Thus, oak stands that are going to be maintained in a forest condition should not be considered for bobwhite management. Although oak woodlands may receive some use by bob-white, landowners managing property specifically for bobwhite should consider reducing the basal area in those stands to at least a residual of 30–40 square feet. The best use of fire in woodlands to enhance conditions for bobwhite is where woodlands are adjacent to high-quality early successional cover, which should dominate any property managed for bobwhite.

Fire is the tool of choice to maintain high-quality early successional cover for bobwhite. Both dormant- and growing-season fire should be used to manage oak savannahs. Relatively short fire-return intervals (2–3 years) are necessary to create good quail habitat in oak savannahs. Where woody stem density needs to be reduced, late growing-season fire should be implemented. Where bobwhite is the focal species, it is critical to prescribe fire on a different area of the property every year, setting back succession and providing diverse structure and composition, both spatially (across the property and landscape) and temporally (different times of the year).

Forest songbirds

There are many species of songbirds that use oak forests. All fill a different ecological niche and have various habitat requirements. Some nest and forage in the overstory, some in the midstory, and some in the understory. Most nest among vegetation, but some nest on the ground.

Burning closed-canopy forests generally will not improve conditions for forest songbirds, especially those that use a well-developed understory, such as black-throated blue warbler, hooded warbler, worm-eating warbler, white-eyed vireo, and Kentucky warbler, or those that require leaf litter for nesting, such as ovenbird. However, forests with a broken canopy that allow at least 20%–30% sunlight to reach the forest floor may be managed with fire to develop suitable structure. Low-intensity dormant- and early growing-season fire on a 3–5-year fire-return interval will maintain a well-developed understory. Relatively open woodlands maintained with a 3–5-year fire-return interval can be managed for the northern flicker, red-headed woodpecker, American redstart, fox sparrow, white-crowned sparrow, and chipping sparrow. Late growing season fire may be used intermittently with dormant-season fire to decrease woody composition and increase vegetation and structural diversity for various species using oak woodlands. For species that use young stands with high stem densities, such as chestnut-sided warbler, dormant-season fire could be used every 5–7 years on sites where the overstory has been removed or where less than 30 square feet of basal area remain to maintain high stem densities.

To avoid disrupting nests, burning should be completed by mid-April where forest songbirds are a consideration. Many forest songbird species nest near the ground among understory vegetation.

Old-field and shrubland songbirds

Many species of songbirds require early successional or old-field vegetation that is provided in oak savannahs. Indigo bunting, field sparrow, dickcissel, blue grosbeak, and others require open areas dominated by forbs and grasses with scattered shrub cover. Savannahs can be burned on a 1–3-year fire-return interval to provide this type of cover and structure. Late growing-season fire may be used intermittently with dormant-season fire to reduce woody composition where needed and increase vegetation and structural diversity.

Other species, such as eastern towhee, gray catbird, brown thrasher, and white-eyed vireo, require more shrub cover for nesting, foraging, and loafing. Areas within woodlands and savannahs may be burned every 3–5 years using dormant-season fire to maintain habitat for these species. Relatively open woodlands with considerable sapling structure will attract more shrubland songbirds than woodlands with a greater basal area of overstory trees.

Reptiles and amphibians

Several snakes and lizards, such as black rat snakes and fence lizards, selectively use recently burned areas in oak systems, especially woodlands and savannahs that are more open. Not only are these species able to find adequate food resources (such as invertebrates and small mammals), they also are better able to absorb necessary heat for thermodynamics in areas that have been burned recently.

Both dormant- and late growing-season fire can be used to enhance habitat for reptiles. However, burning should be completed before these animals leave their hibernacula in the early spring. Burning at that time can cause excessive mortality for these animals as many individuals may be concentrated in a relatively small area and they are more vulnerable to fire because they are out during the daytime trying to warm up and they are not as mobile as they are later when temperatures are warmer. Burning every 2–5 years maintains suitable structure for various reptiles in forests with a broken canopy, as well as woodlands and savannahs.

Most amphibians are associated with wetlands and mesic uplands. Exceptions include the American and Fowler's toads. There are also several species of salamanders found in upland oak forests and woodlands.

Habitat conditions following prescribed fire generally are not favorable for Plethodontid salamanders that live in the leaf litter within a year or so after burning because when the leaf litter is consumed, moist conditions necessary for oxygen exchange through their skin are no longer present. Thus, where fire is going to be used to meet other objectives, but these salamanders still are a concern, a longer fire-return interval and smaller burn units might be considered if possible. For other salamanders that live primarily belowground except during the breeding season, timing of fire is most important. The Ambystomatid salamanders are most active in the late winter as they leave hibernacula and travel to ephemeral ponds where they breed. Burning should be conducted outside this period where these salamanders are a concern (Figure 9.5).

Livestock

Although not commonly practiced anymore in the Eastern United States, woodlands and savannahs can provide range for livestock. Historically, cattle produced in the Eastern United States were commonly supported, especially during the summer months, on open range that consisted of woods that were burned annually. For contemporary woodlands and savannahs to provide valuable livestock forage, several conditions must be met. First and foremost, the overstory canopy must be at a low

Figure 9.5 Low-intensity burns that are patchy may be good for several wildlife species as a mosaic of unburned areas can provide thicker cover for foraging, escape, or nesting. (Photo courtesy of C. Harper.)

enough density (about one-third that of a closed-canopy forest) to foster the growth of an herbaceous ground layer. Second, the woody midstory must be reduced, virtually eliminated, because midstories can suppress the ground layer as effectively as the overstory if they are allowed to become too dense. Finally, the ground layer must have a substantial component of grasses, and they should be desirable forage-producing species. Fortunately, bluestems and other desirable forage species respond well to savannah restoration practices.

Frequent burning on a 1–3-year fire-return interval will be necessary to maintain forage and prevent a midstory from developing. Where woody groundcover is excessive, the site may provide beneficial range for goats. Regardless of the type of livestock used, grazing has always been a natural part of the disturbance that regulated North American woodlands and savannahs. Thus, cattle can be a part of an ecosystem restoration effort or even some wildlife management programs as long as grazing is well managed.

Implementing prescribed fire in oak systems

It is beyond the scope of this chapter to provide a step-by-step prescription for implementing a prescribed fire. However, there are several steps that should be highlighted and followed carefully. Before implementing any prescribed fire, it is critical to work with experienced personnel until you have adequate experience to conduct the burn yourself. And then, it is always necessary to have adequate help to complete the burn (Figure 9.6).

Figure 9.6 Adequate personnel and good firebreaks are essential when implementing prescribed fire. (Photo courtesy of C. Harper.)

Preparing the site

Before any site can be burned, there must be sufficient firebreaks. A creek or road can serve as a firebreak, but usually one has to be established. A dozer and fire-plow is often used by state forestry agencies. These firebreaks can be made more permanent by dressing them afterwards with a tractor and disk to smooth out the firebreak, which makes it easier to walk or ride along with an ATV. Access can be important when setting backing fires and checking the area after burning. If adequate sunlight is available, firebreaks can be planted if desired. Of course, firebreaks should be established along contours with consideration to prevent erosion and sedimentation. Firebreaks in woods can also be established with backpack blowers. Clearing leaf litter is easily accomplished by a couple of people with backpack blowers, and there is no soil disturbance and topography is less an issue than with dozers or tractors. Any standing dead trees within reach of the firebreak should be felled to prevent them from falling if they begin to burn and spreading embers across the firebreak. Although it is best to establish a firebreak that exposes soil (which does not burn), firebreaks along the edges of woods and savannahs can be created with water trucks that have the capacity to spread enough water. Volunteer fire departments often have this capability and may be willing to help implement the prescribed fire. *It is important to realize mowed strips of vegetation do not stop fire and should not be considered a firebreak unless adequate water is used!*

The burn plan

A burn plan should be prepared prior to every burn. Most state forestry agencies provide burn plan templates online. The burn plan should describe the area to be burned, state the objectives, and list preburn factors, such as a description of the fuels present, manpower and equipment needed, nearby smoke management considerations, and ignition procedure. A map of the area to be burned should be included in the plan. The map should clearly show the firebreaks and the planned ignition procedure. The desired, predicted, and actual weather conditions should be included in the plan. Obviously, actual weather conditions will not be known until it is time to burn; therefore, the burn plan is a working document that is filled out before, during, and after the burn. Along with actual weather conditions, the desired and actual fire behaviors should be recorded as well. Finally, after the burn, a post-burn evaluation should be completed. All this information will help ensure you are prepared to implement the fire and help you understand the effects of the burn after completion. In summary, this helps make you a more successful and safe prescribed fire manager.

Notifying appropriate contacts prior to burning

A critical step prior to burning is notifying the proper contacts. In most states, it is necessary to obtain a burning permit prior to burning, at least during certain seasons of the year. In addition to this, you should notify 911 and tell the dispatcher that you have a burn permit and that you intend to burn during the stated time of day. You then should call the area fire department(s) and tell them you have a burn permit and that you intend to burn during the stated time of day. Do not hesitate to ask them to come out and help if they would like. Often, a local volunteer fire department enjoys coming out and helping. Involve everyone you can. Make sure your neighbors know what you are doing! The more contacts you make, the better. No one in the local area should be surprised when they see smoke.

DO NOT BURN DURING "RED FLAG" CONDITIONS

You have planned the burn for months. You have everything in order and plenty of people on-site ready to help. Everything has come together, except.... In your eagerness to burn, do not overlook a "red flag" condition and think "It will be alright." That is a recipe for disaster. Do not hesitate to cancel a burn if conditions are not right, have changed, or if something has been overlooked. There are several "red flag" conditions and you should not burn if any are present.

Do not burn if

- Adequate firebreaks are not in place
- You have not completed a written burn plan
- Everyone has not been briefed on the plan and conditions
- Conditions are too dry or too windy
- The forecast does not meet the prescription
- There is inadequate personnel or if experienced personnel are not present
- Communication (such as hand-held radios) is not available for everyone helping
- Adequate contacts have not been notified

Firing techniques

There are five main firing techniques commonly used when prescribing fire. The recommended firing technique varies with landowner objectives, condition of fuels, topography, and weather. Different firing techniques allow you to adjust fire intensity to the desired level. Although it is beyond the scope of this chapter to provide a detailed description of each firing technique and when they should be used, a general description is provided to introduce you to firing techniques that may be used when burning oak systems. For additional information, contact your state forestry agency and review more detailed publications on using prescribed fire, such as Waldrop and Goodrick (2012) and Weir (2009).

Backing fire

A backing fire moves against the wind. Therefore, it is relatively slow moving and generally consumes fuel more completely than fire moving with or parallel to the wind. Backing fires are set along a firebreak (or other barrier, such as a creek or road) on the downwind side of the area to be burned and allowed to back into the wind. Steady wind speed and direction are important for consistent and predictable burning. Backing fire is less intensive with lower flame lengths than other firing techniques and produces less scorch (if any) when burning oak woods. However, the slower moving fire can have a longer residence time and can possibly damage fine feeder roots if the duff layer below the leaf litter is dry. Backing fire should be used when considerable fuels are present, especially vertical fuels (dense shrub layer), as they can elevate fire intensity if a heading fire is used, which may damage or kill desirable overstory trees. Backing fire typically is the easiest to implement. Although it takes longer to implement, a backing fire is the recommended firing technique when burning oak woods where damage to overstory trees is not desirable.

Strip-heading fire

A strip-heading fire involves both backing fire and heading fire. A backing fire is first set along a firebreak; then firelines are set sequentially upwind of the backing fire and allowed to move into the backing line of fire. The distance of the strip-heading firelines from the backing fireline is set according to the desired level of fire intensity. Normally, the strip-heading fireline is set 50–200 feet upwind of the backing fire. Fuel conditions, wind speed, relative humidity, and landowner objectives determine the distance that should be used. A major advantage of strip-heading fires over backing fires is strip-heading fires require much less time, allowing quick ignition and burnout. A major consideration when using strip-heading fires is to create an adequate blackline (at least twice the distance that will be used between the backing fire and the strip-heading fireline) with the initial backing fire prior to setting the first strip-heading fireline. When burning oak woods, strip-heading fires are applicable when relatively flat fuels are present. Presence of vertical fuels may demand increased use of backing fires.

Flanking fire

Flanking fires are set directly into the wind, allowing the fire to burn at right angles to the wind. Obviously, this can be unpredictable and dangerous if wind direction is inconsistent. Typically, several burners are necessary to implement a flanking fire. Coordination is critical so that all the firelines are set at the same time and all the burners know where each of them are and move at the same speed. Flanking fire is also used to secure the flanks of backing and strip-heading fires along firebreaks. Flanking fire usually is intermediate in intensity between backing and strip-heading fires. Implementation is more difficult, however, as coordination is critical among burners.

Point-source fires

Point-source fires can be viewed as modified strip-heading fires, where spot ignition is used instead of strip ignition. Spot ignition produces more intensity than the backing fire, but less intensity than strip ignition. This can help speed up the line-backing fire with less intensity than would be produced with a strip-heading fire. Point-source ignition spots are ignited upwind of the backing fireline. The distance of the point-source ignition spots from the backing fireline is determined just as those in a strip-heading fire. It is important to realize point-source ignitions spots are not placed randomly. To safeguard against hot spots, point-source ignition spots should be equidistant along each fireline. If not equidistant, a point-source ignition spot on one fireline may burn between two point-source ignition spots on the adjacent downwind fireline and produce a heading fire that exceeds the desired intensity levels. Firelines may be set by a single burner, or firelines may be set simultaneously by multiple burners. If multiple burners set firelines, it is critical that they are able to communicate (they should be able to see each other and communicate via radio) and are coordinated in setting the firelines.

Ringfires or center fires

A ringfire is ignited by first igniting the backing fire adjacent to the downwind firebreak. After securing the firebreak, the fire is spread around the perimeter of the area being burned, along the firebreak that surrounds the area. Before the perimeter is complete, a point-source fire is ignited in the center of the area. Then, the remainder of the perimeter is ignited. Convection from the center point-source fire creates indrafts that pull the perimeter fire inward, potentially creating a strong convection column, which can cause spot fires a considerable distance away. Ringfires are normally used only when clearing debris prior to planting trees. It is described here only for informative purposes. We do not recommend ringfires when burning in oak systems. Furthermore, ringfires should not be used whenever wildlife is a consideration as direct mortality is most likely with this firing technique.

Smoke management considerations

The old adage "where there is smoke, there is fire" can be flipped to "where there is fire, there is smoke." And when we create smoke, just like fire, we are responsible, and we must consider potential impacts of smoke before implementing a fire. Major considerations include atmospheric stability, wind direction, fuel moisture, weather forecast, and timing.

Controlled burning should not be implemented when the atmosphere is stable and smoke is unable to rise and disperse. It is most desirable for smoke to rise rapidly, disperse, and be transported away from the site relatively quickly. Winds should carry smoke away from sensitive areas, such as major roads, hospitals, airports, and schools. Burning should not be conducted at any time that smoke may affect such sensitive areas. These considerations do not prevent prescribed fire from being used near sensitive areas. However, wind direction and atmospheric stability must be considered closely so smoke is moved away from those sites and dissipates relatively quickly.

Adequate moisture should be available to prevent consuming the duff layer under the leaf litter when burning oak woods. However, excessive moisture can lead to smoke management problems. Trying to burn when relative humidity is above 50% can lead to excessive smoke. Burning during the growing season will produce more smoke than burning during the dormant season because of the moisture in the green leaves in the understory. Balancing fuel moisture with smoke management considerations is important.

The amount of smoke produced is increased with higher relative humidity and moisture conditions. Burning during the middle of the day typically produces less smoke than burning during the early morning or evening because relative humidity is lower and the ambient temperature is greater, which allows more complete fuel consumption and less smoke. As temperatures decrease in the evening and relative humidity rises, smoke problems can be accentuated, especially in low-lying areas. This situation can be particularly problematic if smoke follows drainages and meets a road. Also at this time, wind direction is more difficult to predict, and winds often cease completely, allowing smoke to settle and reduce visibility. This is not to say burning should not be conducted during morning, evening, or nighttime. Burning during these times may be most sensible if conditions are best to meet management objectives. For example, when overly dry conditions prevail, burning during midday may not be prudent because fire intensity may be too great.

Small test fires should be set prior to initiating a prescribed fire to confirm predicted fire intensity, wind direction, fuel consumption, and smoke dispersion. Contingency plans should be in place if conditions change unexpectedly and burning should be halted.

Post-burn evaluation

Evaluating the results of your management efforts is critical to success. Burning just because someone recommended does not make sense, especially if the desired results or effects are not realized. In many cases, you will not know if the desired results are achieved unless you evaluate your efforts. Post-burn evaluation should be completed immediately or soon after a burn, as well as during the growing season following the burn.

Surveying the area immediately after a burn allows you to record the percentage of the area that burned and evaluate fire intensity. If some areas did not burn, determine why they did not burn, and note whether that is desirable or not. If not, what conditions were present that kept that area(s) from burning. Fire intensity can be evaluated by recording how much of the litter layer and understory vegetation was consumed and if any desirable overstory trees were damaged. If there were any damaged, was it because the fire was too intense, or because debris was present around the base of the tree? If a growing-season fire was implemented, was the understory foliage consumed, or was it just wilted and discolored? Did the duff layer remain intact? Recording this information will help you determine the intensity of fire under the present conditions, and if your next burn should be altered depending on the results after burning in those conditions. In your burn plan, there should be a place to record this information, as well as other observations, such as characteristics of smoke dispersal, flame lengths, how long it took to complete the burn, and incidence and timing of smoldering. Incidence of fire escape should be recorded. If the fire escaped the firebreak, you should know what action should be taken next time to prevent a similar response. Wildlife mortality should be recorded. For those who have not burned before, they often are interested and pleasantly surprised to find no wildlife harmed by the fire. If you do find wildlife mortality, consider your timing of burning and the firing technique used and how that might be adapted next time to prevent mortality (Figure 9.7).

In the growing season following the burn, continue to evaluate the fire by recording the understory vegetation response. Is there an increase in herbaceous plants, or are woody sprouts dominant? Are woody seedlings present? What percentage of woody sprouts top-killed by the fire is re-sprouting? Is damage to any overstory trees obvious? Recording this information will help you make any necessary adjustments to burning technique, timing, and intensity.

Figure 9.7 It is important to evaluate the site after burning to estimate the percent of the area burned, amount of leaf litter and other fuels consumed, and to survey the area for mortality of wildlife, such as snakes, turtles, and rabbits. (Photo courtesy of C. Harper.)

Human perception: Opportunities and limitations

As mentioned earlier in this chapter, our society has been convinced all fires in the woods are bad. The notion that any fire is bad and that it "destroys" the forest is ridiculous. Fire is a natural and necessary phenomenon. Fire causes change. By using prescribed fire accordingly, you can make sure the change is directed toward helping you meet your land management objectives.

The best way to change human perception of using fire in the woods is to use it and show the effects. Invite others to come see the effects of your woods burning efforts. Invite them to come help in your next controlled burn. Always remember to make appropriate contacts before burning, and never allow your neighbors to be surprised when you burn. Over time, do not become too "comfortable" when burning. This leads to laziness, taking risks, and not making sure all precautions are met. Burning should never be conducted outside the recommended prescription.

When the appropriate precautions and procedures are taken and followed, fire is a great tool to help meet management goals and objectives in oak systems, whether they focus on wildlife considerations, aesthetics, forest health, or timber production. Keeping an open mind with regard to possible management practices, and adjusting techniques as necessary to meet your objectives will help you enjoy your forest resources more than ever.

Further suggested reading

Brose, P.H., D.C. Dey, T.A. Waldrop. 2014. The fire–oak literature of eastern North America: synthesis and guidelines. USDA Forest Service General Technical Report NRS-135, Newtown Square, PA.

Cohen, D., B. Dellinger, R. Klein, and B. Buchanan. 2007. Patterns in lightning-caused fires at Great Smoky Mountains National Park. *Fire Ecology* 3(2):68–82.

Delcourt, H.R. and P.A. Delcourt. 1997. Pre-Columbian Native American use of fire on southern Appalachian landscapes. *Conservation Biology* 11:1010–1014.

Dey, D.C. and C.J. Schweitzer. 2015. Timing fire to minimize damage in managing oak ecosystems. Proceedings of the 17th biennial southern silvicultural research conference. USDA Forest Service General Technical Report SRS–203. Asheville, NC.

Donovan, G.H. and T.C. Brown. 2007. Be careful what you wish for: The legacy of Smokey Bear. *Frontiers in Ecology and the Environment* 5(2):73–79.

Flatley, W.T., C.W. Lafon, H.D. Grissino-Mayer, and L.B. LaForest. 2013. Fire history, related to climate and land use in three southern Appalachian landscapes in the eastern United States. *Ecological Applications* 23:1250–1266.

Frost, C.C. 1998. Presettlement fire frequency regimes of the United States: A first approximation. *Proceedings of the Tall Timbers Fire Ecology Conference* 20:70–81.

Holoubek, N.S. and W.E. Jensen. 2015. Avian occupancy varies with habitat structure in oak savanna of the south-central United States. *Journal of Wildlife Management* 79:458–468.

Johnson, A.S. and P.E. Hale. 2002. The historical foundations of prescribed burning for wildlife: A Southeastern perspective. In W.M. Ford, K.R. Russell, and C.E. Moorman (eds.), *The Role of Fire in Nongame Wildlife Management and Community Restoration: Traditional Uses and New Directions.* USDA Forest Service General Technical Report NE-288, Newtown Square, PA, pp. 11–23.

Kabrick, J.M., D.C. Dey, C.O. Kinkead, B.O. Knapp, M. Leahy, M.G. Olson, M.C. Stambaugh, and A.P. Stevenson. 2014. Silvicultural considerations for managing fire-dependent oak woodland ecosystems. USDA Forest Service GTR-NRS-P-142.

Knapp, B.O., K. Stephan, and J.A. Hubbart. 2015. Structure and composition of an oak-hickory forest after over 60 years of repeated prescribed burning in Missouri, U.S.A. *Forest Ecology and Management* 344:95–109.

Knapp, E.E., B.L. Estes, and C.N. Skinner. 2009. Ecological effects of prescribed fire season: a literature review and synthesis for managers. Gen. Tech. Rep. PSW-GTR-224. Albany, CA: U.S. Department of Agriculture, Forest Service, Pacific Southwest Research Station, 80 pp.

Lashley, M.A., C.A. Harper, G.E. Bates, and P.D. Keyser. 2011. Forage availability for white-tailed deer following silvicultural treatments in hardwood forests. *Journal of Wildlife Management* 75:1467–1476.

McCord, J.M., C.A. Harper, and C.H. Greenberg. 2014. Brood cover and food resources for wild turkeys following silvicultural treatments in mature upland hardwoods. *Wildlife Society Bulletin* 38:265–272.

Nowacki, G.J. and M.D Abrams. 2008. The demise of fire and "mesophication" of forests in the eastern United States. *BioScience* 58:123–138.

Peterson, S.M. and P.B. Drewa. 2006. Did lightning-initiated growing season fires characterize oak-dominated ecosystems of southern Ohio? *Journal of the Torrey Botanical Society* 133:217–224.

Pinchot, G. and W.W. Ashe. 1897. *Timber Trees and Forests of North Carolina*, Bulletin No. 6. North Carolina Geological Survey, Raleigh, NC, 227 pp.

Robertson, K.M. and T.L. Hmielowski. 2014. Effects of fire frequency and season on resprouting of woody plants in southeastern U.S. pine-grassland communities. *Oecologia* 174:765–776.

Russell, K.R., D.H. Van Lear, and D.C. Guynn. 1999. Prescribed fire effects on herpetofauna: review and management implications. *Wildlife Society Bulletin* 27:374–384.

Sutherland, E.K. 1997. History of fire in a southern Ohio second-growth mixed-oak forest. In *Central Hardwood Conference*, Vol. 11. USDA Forest Service General Technical Report NC-188, St. Paul, MN, pp. 172–183.

Van Lear, D.H. and T.A. Waldrop. 1989. History, uses, and effects of fire in the Appalachians. USDA Forest Service General Technical Report SE-54.

Waldrop, T.A. and S.L. Goodrick. 2012. Introduction to prescribed fire in Southern ecosystems, Science Update SRS-054. U.S. Department of Agriculture Forest Service, Southern Research Station, Asheville, NC, 80 pp.

Weir, J.R. 2009. *Conducting Prescribed Fires: A Comprehensive Manual.* Texas A&M University Press, College Station, TX, 194 pp.

chapter ten

Competition control for managing oak forests

Victor L. Ford, Jim H. Miller, and Andrew W. Ezell

Contents

Principles of competition control.. 132
General plantation establishment prescription 133
Site preparation for oak planting after complete or partial harvest 133
    Mechanical treatments.. 133
    Prescribed burning.. 135
    Herbicide treatments 136
        About herbicides... 136
        Herbicide labels 138
        Timing of herbicide applications.. 138
Woody plant control with tree injectors and backpack sprayers 139
    Stem injection.. 139
    Cut-treat...141
    Directed foliar sprays and wipes 143
    Basal sprays and wipes.. 143
Broadcast herbicide applications 144
    Utility skid and trailer-mounted sprayers................................. 144
    All-terrain vehicle and recreational-type vehicle-mounted
    sprayers... 145
    Tractor-mounted sprayers... 145
    Backpack mist blowers 145
    Aerial sprayers.. 145
Site preparation on old-field sites... 146
Site preparation 146
Herbaceous weed control 147
Final thoughts... 151

Other than site selection, competition control is the most critical factor in the survival and growth of oak plantations. In natural stand development, there are thousands of oaks competing with other plants and each other to dominate the stand. Plantations use only a comparatively few individuals and give them the opportunity to dominate by controlling other vegetation. The trade-off is natural stands require decades to capture the site, but plantations can do this in a few years. Competition control in natural stands can decrease the time to maturity and fruiting from a couple of decades to a single decade. The key is to give each individual oak an opportunity to capture the site with minimal interference.

Principles of competition control

Over the past few decades, several principles of vegetation management for plantation establishment have been discovered. These principles as they apply to oak culture can be summarized as follows:

- Early growth gains from years one through three are maintained and amplified well into the rotation.
- Vegetation control can increase tree growth by the greatest proportion in years one through three, especially when near complete control is achieved around large, properly planted, healthy seedlings that are not injured by the vegetation control treatments.
- Threshold levels of competition reduction must be achieved before a positive growth response occurs and a point of diminishing returns also exists where additional control gains no further growth response.
- Plants immediately surrounding crop tree seedlings are the strongest competitors and should be the main target for control.
- A mixture of grasses, ferns, forbs, vines, and woody sprouts present severe competition to tree seedlings—grasses and ferns are most competitive in the early stages from years one through five due to full sunlight and the ability to capture the site in a single year.
- Woody plants become established in greatest numbers during the first year after clearcut harvesting from vigorous sprouting; species such as yellow poplar, black cherry, and fire cherry will reproduce from seedlings; preestablished oaks will often become established from seedling sprouts.
- The most severe woody competitors are not shrubs, but trees, which can grow equally with crop trees and maintain a position in the main canopy; re-sprouting hardwoods are the most severe woody competitors to planted seedlings.
- Following mechanical site preparation, woody competition starts exerting its influence in years five through eight, but woody control is most cost effective prior to planting or in years one and two.

- Tree diameters respond proportionally more than heights to competition control.
- Projected returns on investments in intensive site preparation treatments are usually only justified on sites of medium to high quality.

Beyond competition control, oak seedlings benefit from soil tillage (both surface and subsoil) presumably by aiding rapid root development. It is also assumed site preparation requirements will probably be more stringent for oak seeding than for seedling planting. These principles and concepts require further verification for specific oak species as they interact with the multitude of sites and plant community situations.

General plantation establishment prescription

A general prescription for oak plantation establishment is to plant large, healthy seedlings on tilled and appropriately subsoiled sites with minimum hardwood competition, and herbaceous vegetation control treatments applied annually for 2–3 years. Both woody and herbaceous control treatments must be effective to realize a return on the investment. Because of the constant immigration of hardwoods into young stands, woody control treatments should be performed at regular intervals as needed. Rapid early seedling growth is critical to oak establishment to minimize predation from deer, rabbits, and rodents. Only through enhanced early growth can time spent in vulnerable stages be shortened. At the same time, site preparation treatments must work in concert with the overall management plan and safeguard the multiresource values of the forest, maintain or improve long-term soil productivity, and protect intrinsic site values.

Site preparation for oak planting after complete or partial harvest

Intensive site preparation treatments in preparation for oak planting on cutover forest sites often require the integrated use of mechanical, burning, and herbicide treatments. Each of these tools has implications for wildlife habitat quality. These issues are discussed in some detail in the following sections.

Mechanical treatments

Mechanical competition control is any method using physical means to reduce the advantages of the competition. Plowing, disking, and hoeing are all mechanical methods traditionally used to control competing vegetation. While there is no danger of herbicide damage or offsite movement with mechanical methods, they are not without potential problems.

Mechanical methods under certain conditions can lead to erosion and degradation of water quality.

The advantages of using mechanical site preparation treatments before planting or seeding oaks are numerous. Logging slash or remaining noncrop trees and shrubs can be felled and pushed into piles or windrows using shear blades on bulldozers and root raking, or processed with choppers. Re-sprouting vigor of hardwoods can be decreased significantly by felling or shearing operations after the spring growth flush has depleted root reserves. Root raking is also used to dislodge woody plant roots, minimizing woody competition and increasing available nutrients and water for crop-tree growth. Clearing the nonharvested trees and logging debris permits other soil amelioration treatments, such as disking (harrowing) and bedding. Any mechanical disturbance that sets succession back to bare-soil conditions can provide excellent habitat for species associated with such seral stages, the northern bobwhite being an excellent example. However, the herbaceous plant communities that develop after these disturbances can also be valuable as brood areas for wild turkey and ruffed grouse. Windrowing and the heavy cover it creates can also provide nesting cover and excellent escape cover for species such as cottontail rabbits. On the other hand, such cover can also be beneficial for various predators that can negatively impact ground-nesting game birds and rabbits.

On upland and poorly drained sites, disking can benefit oak establishment. On poorly drained sites, bedding has yielded short-term growth increases and considerable increases in survival. Disking increases the rootability of the surface soil across the entire area, while bedding does the same and locally raises the soil above the winter water table. This increase in the water table generally is the result of the removal of the overstory. Bedding allows the seedlings to survive until they can reduce the water table. However, the volume of tilled soil is less with bedding compared to disking, and root closure between rows may not occur as quickly as on disked sites. On wetter sites, bedding not only creates drier areas (the beds themselves) but also provides areas that are wetter than they would have otherwise been (the furrows). These furrows can often hold water well into the growing season and as such can be valuable as breeding areas for various amphibians such as frogs, toads, and pond-breeding salamanders.

The principal disadvantage of root-raking treatments occurs when valuable site nutrients are displaced into windrows and piles, and away from planting rows and seedling access. Also, excessive erosion can occur, especially as slopes increase and terrain is variable. Further, raked or disked organic matter decomposes faster, perhaps before seedling roots can take full advantage of the ensuing nutrient release. Soil disturbance also clears the organic mulch that aids in preventing evaporative moisture loss from the soil surface. At the same time, an abundance of herbaceous plants can become established on the bare soil, resulting in severe early competition to planted oaks. The herbaceous community also provides excellent habitat for rodents that encourages seed and seedling predation. However, on some sites, revegetation after intensive mechanical scarification can be slowed or spotty, leading to erosion and nutrient loss. Only through careful and well-considered application of mechanical treatments can these adverse effects be minimized or eliminated. Combination plows, also known as "3-in-1 blades," can disc, subsoil, and bed the site in one pass and have been used to successfully establish oaks in the South.

As a stand gets older, mowing (bush-hogging) and cultivation can be employed as mechanical competition control practices. These methods must be repeated multiple times each year for multiple years to be effective. You can expect between 2% and 12% mortality of crop trees with each mechanical treatment. Planning and implementation of correct row spacing facilitates cultivating between rows and cross rows. Without this, cultivation is difficult, if not impossible, and options are greatly diminished. In such cases, hoeing may be the only option. Mechanical control is generally more expensive than chemical methods and usually not as effective; weeds generally recapture the site before the oaks.

Prescribed burning

Prescribed burning is commonly used after mechanical and chemical site preparation to further reduce logging debris and thus improve planter access. Prescribed fire can, with adequate fuel and proper burning conditions (timing), top-kill woody plants less than 3 inches groundline diameter but increases herbaceous groundcover. This increase in groundcover may provide excellent wildlife forage. Fire has many impacts on litter depth, woody debris loads, seed scarification, soil nutrition dynamics, plant species composition, and structure, all of which have important implications for wildlife habitat quality. The implications of fire on wildlife habitat in eastern oak forests are treated in much greater detail in Chapter 9.

The effectiveness of a prescribed burn depends on the intensity and timing relative to target plant size and development. Fire intensity is determined by the amount of fuel and its arrangement and dryness, along with weather, topography, and ignition source and pattern (see Chapter 9).

The most effective time for weakening woody plant competitors is burning in the summer after plants have initiated growth and used their root reserves. Burning in the late winter or spring leaf-out can minimize the period of bare soil, while summer burns maximize consumption of logging debris and are more effective at controlling competition. Burning can, however, predispose a forest stand or opening to nonnative plant invasion. A close evaluation of the benefits and risks is warranted before applying prescribed fire.

Herbicide treatments

The most effective method of competition control is the use of herbicides. Application of herbicides can suppress competing vegetation longer than mechanical methods. Herbicides can be used by broadcast and selective applications for site preparation before planting and by selective applications for herbaceous and woody plant control after planting.

About herbicides

Modern herbicides are both safe to the applicator, wildlife, and the environment when used according to label instructions. State applicator certification programs and strict adherence to the label requirements during transport, mixing, application, and container disposal will assure safe use. Consult with the Extension Service at your land grant university for further information on labels and specific information, such as herbicide rates, prohibitions of use, and disposal requirements for herbicide containers. It is critical that you select the most effective herbicide or tank mix for the target species as well as for the constraints of the site itself.

Herbicides should be (1) mixed thoroughly in clean water with the proper additives, (2) applied correctly to minimize nontarget damage and offsite movement, and (3) applied when they will be most effective. The herbicide label will give specific details on all these points and should be thoroughly read and understood before purchase or application. Likewise, you should understand and follow precautions that apply to re-entry of the site after application. Personal protective equipment that must be used and maintained by applicators and loaders will be specified on the label.

Herbicides are identified by brand name and common name. The brand name is specific to the manufacturer and generally capitalized. The common name refers to the active ingredient. Because most of the compounds have several manufacturers, we will use common name and a brand name example in parenthesis. The authors are not endorsing any particular brand. Soil active herbicides should be used with a full appreciation of the risk of nontarget plant damage as a result of residual carryover or application proximity. But it also should be recognized that herbicides having both foliar and soil activity are often the most effective. Those herbicides discussed in the following sections are described further in Table 10.1.

Table 10.1 Herbicides labeled or used for oak culture, manufacturers, active ingredients, and concentrations in formulations

Product	Manufacturer	Active ingredient (s)	Amount of a.i. or a.e.[a] ingredient(s) in formulation
AAtrex 4L[b]	Syngenta	atrazine	4 lb/gal
Access[b]	DowAgroSciences	picloram + triclopyr	1 + 2 lb/gal
Accord, Roundup, generics	Several	glyphosate	4 lb/gal
Arsenal AC	BASF	imazapyr	4 lb/gal
Atrazine 4L[b]	Several	atrazine	4 lb/gal
Chopper	BASF	imazapyr	2 lb/gal
Chopper RTU	BASF	imazapyr	3.6%
Escort XRT	Du Pont	metsulfuron	60%
Fusilade	Syngenta	fluazifop	1 lb/gal
Garlon 3A	DowAgroSciences	triclopyr amine	3 lb/gal
Garlon 4	DowAgroSciences	triclopyr ester	4 lb/gal
Oust, generics	Du Pont, Several	sulfometuron	75%
Pathway, Tordon 101R and RTU	DowAgroSciences	2,4-D + picloram	¼ + 1 lb/gal
Princep 4L[c]	Syngenta	simazine	4 lb/gal
Tordon K[b]	DowAgroSciences	picloram	2 lb/gal
Tordon 101[b]	DowAgroSciences	2,4-D + picloram	½ + 2 lb/gal
Vantage, Select	Several	sethoxydim	1 lb/gal
Weedone 2,4-DP	NuFarm	2,4-DP amine	4 lb/gal
2,4-D ester	Several	2,4-D ester	4 lb/gal

[a] a.i, active ingredient; a.e., acid equivalent.
[b] Restricted use herbicides that must be applied by a state-certified applicator or permitted private landowner (contact county agent for permit).
[c] Other formulations of simazine are Princep Caliber 90, Princep 4G, and Princep 80W.

HERBICIDES AND WILDLIFE

With respect to wildlife, modern herbicides should be considered safe—when used properly. These chemicals have very limited direct toxicity and the volume of active ingredient applied on a per acre basis is extremely low. As an example, table salt is about 4.4 times more toxic than triclopyr, a widely used herbicide in oak management. A common application rate for this herbicide, 9 lb active ingredient per acre, works out to be 0.00146 oz at the scale of a normal 9″ dinner plate—or about 1/18th of the amount of salt in one of those small white packets you get at the restaurant (0.026 oz).

So when you pour one of those small packets on your meal at the next picnic, just remember that you are putting out a substance that is 4.4 times more toxic and at 18 times the rate (18 × 4.4 = 79.2 times more toxic) of a forestry herbicide! And that herbicide will be applied about once or twice in 50 years versus the daily use of that salt.

The real impact of herbicides on wildlife is through their effect on vegetative structure and composition. In managing competition in oaks, you will generally be trying to kill either competing woody species (e.g., red maple, black locust, or sweetgum) or herbaceous plants (e.g., tall fescue, goldenrod, serecia lespedeza). The issue then is what will replace those competitors that you are eliminating and what is the relative value of each? Also, if you are helping ensure the next oak forest, short-term impacts have to be weighed against long-term benefits. So is the value of a sweetgum sprout over a year or two greater or less than a liberated oak seedling that may endure for more than a century? Where you are controlling undesirable vegetation (e.g., tall fescue, serecia lespedeza, red maple) there really is no downside. Keep in mind that measurable impacts of herbicides in most cases disappear after 1–3 growing seasons—they are short-lived, especially in herbaceous plant communities.

Herbicide labels

The herbicide label is a legal document that specifies on what types of sites and how an herbicide can be applied. Herbicides legally used in forestry must be labeled for "forest sites," or in some instances, for "noncrop areas," "wildlife habitat," "wildlife openings," and "tree farms" when not broadcast for site preparation. The use of herbicides in oak establishment that do not have one of these site specifications could be unlawful. You should check with your herbicide distributor or state extension specialist regarding your particular site if you are uncertain about the use of a particular herbicide. Other specifications on a herbicide label, such as crop-tree species and target efficacy, relate to product performance, manufacturer liability, agency or company policy or state law (e.g., it is required that the target and/or crop tree are listed on the label). It may be legal to use a herbicide for oak culture when the label states "for conifer release," but the manufacturer is not liable for poor performance or crop injury. Remember, it is unlawful to exceed labeled rates, which can occur when numerous stems are treated individually across a site.

Timing of herbicide applications

Herbicides perform best when applied at times when target plants are most susceptible and/or crop trees are most resistant to injury. Applying them before or after the correct time reduces or even eliminates their effectiveness and may damage the crop. As far as the timing of woody control treatments is concerned, to be most efficient, herbicide applications should only be made after all sprouts have emerged following harvesting, burning, or mechanical disturbance. Woody rootstocks must have sprouts before herbicide activity and control can occur.

Woody plant control with tree injectors and backpack sprayers

Manually applied treatments for woody plant control used for oak establishment are tree injection, stump sprays, directed foliar sprays, and basal bark sprays. Soil spot applications using Velpar L® by Du Pont appear to have limited use in hardwood culture because of the residual nature of Velpar L. All sizes of trees and shrubs can be controlled by using the right treatment when the proper herbicide is applied at the correct time. What follows are the sizes of woody plants that can be treated most effectively by manual application methods.

Method	Effective size of target stems controlled
Stem injection	Sizes greater than 2 inches dbh
Cut-treat stumps	All sizes
Directed foliar sprays	Up to 6 feet tall
Full basal sprays	Up to 6 inches dbh
Streamline basal sprays	Up to 2 inches dbh

A combination of methods can be used on the same site when various target stem sizes are treated. Often on the same site, tree injection is used for the larger trees while basal or foliar sprays are applied to smaller woody competitors, and stumps of harvested trees are sprayed.

Stem injection

Stem injection (including hack-and-squirt) involves herbicide concentrate or herbicide–water mixtures applied into downward incision cuts spaced around woody stems. Cuts are made by ax, hatchet, machete, brush ax, cane knife, or a variety of cutting tools and even cordless drills. Tree injection is a selective method of controlling larger trees, shrubs, and vines (greater than 2 inches in dbh) with minimum damage to surrounding plants. Stem injection is the fastest and most cost-effective method to kill individual trees and large shrubs. Injection treatments are sometimes not as effective in controlling multiple-stemmed species compared to the faster basal bark treatments, but may be easier in remote or rough terrain where a backpack sprayer might be impractical or cumbersome. Stem injection is physically demanding for the applicator, who must repeatedly and accurately strike target trees with a sharp tool before delivering the herbicide into the cut. For best results, sharpen tools frequently.

Incisions must be deep enough to penetrate the bark and inner bark, slightly into the wood. Do not make multiple cuts directly above or below each other because this will inhibit movement of the herbicide within the stem. A complete girdle or frill of the stem is not needed or desirable. Space the injection cuts 1–1½ inches apart edge to edge (or per label instructions) around the circumference of each trunk individually or within a clump at a convenient height. Use a handheld, chemical-resistant 1- to 2-quart spray bottle to apply 0.5–2 ml of concentrated herbicide or dilutions (prescribed on the label) into the cut. The amount will depend on the size of cut and how much the cut can hold without the herbicide running onto the bark. Apply herbicide to each cut until the exposed area is thoroughly wet. The herbicide should remain in the injection cut to avoid wasting herbicide and to prevent damage of surrounding plants. All injected herbicides can reach untreated plants by root grafts between like species and uptake of root exudates by all species. These unintended effects are called *flashbacks*. Formulations with imazapyr, such as Chopper® and Arsenal®, will cross root grafts or be exuded in the soil and taken up by nearby roots of neighboring plants. If crop trees are present on a site, these formulations should not be used. Formulations with glyphosate, such as Accord or Roundup, are not soil active and cannot be taken up by neighboring plants, but can be translocated across root grafts. Root grafts can only occur between the same species or closely related species. If oak crop trees are present, herbicides such as Accord can be used if oaks are not treated. Herbicides with triclopyr do not have the tendency to cross root grafts and can be used without much concern with flashback. Crop trees may be injured with triclopyr but it usually is not fatal. Herbicides with soil activity can damage nearby plants when washed from incisions into the soil by unexpected rainfall soon after application. Avoid injection treatments if rainfall is predicted within 48 hours.

Tree injection can be applied at most times throughout the year, but December through mid-January seems least effective. Prolonged cold temperatures can freeze herbicide in the cut, resulting in poor absorption. Heavy spring sap flow can wash herbicide from incision cuts, resulting in poor control and soil transfer to nontarget plants. Prolonged and severe drought is also an ineffective application period.

Herbicides labeled for tree injection that have wide control spectrums are

- Arsenal AC®
- Accord® and Roundup®
- Chopper

- Garlon 3A®
- Pathway®, Tordon RTU® (ready to use)
- Tordon 101R® and Tordon 101®
- 2,4-D

Some of these herbicides have the same active ingredients with new names or a slight difference in formulation (Table 10.1). Accord is the same as Roundup without a surfactant and the active ingredient is glyphosate. Chopper is half the concentration of imazapyr than found in Arsenal AC. Pathway is the new name for Tordon RTU and Tordon 101R and the active ingredient is picloram. Accord will completely replace Roundup, and Pathway will replace Tordon formulations as the other names are phased out. Efficacies of these herbicides for selected species are presented in Table 10.2. Garlon® (triclopyr) is the preferred herbicide in most herbicide application in crop trees because there is less movement of this herbicide from treated stem to untreated stems than most herbicides.

Garlon 3A and Accord (Roundup) have the advantage of no soil activity. Of these, Garlon 3A is preferred because it is effective on more species, especially maple and hickory, and does not cross root graphs as readily as Accord. Arsenal AC, though soil active, has the broadest spectrum of control of any of these herbicides and can be used at wider spaced injection cuts. Garlon 3A and Arsenal AC are usually applied diluted at 33%–50% and 5%–10%, respectively. All products can be applied year-round, except during times of heavy sap flow in the spring. Arsenal AC and Chopper are most effective when injected from July to October. Arsenal and Chopper applications need at least a year for the herbicide to dissipate before regenerating the site.

Cut-treat

Cut-treat involves applying herbicide concentrates, herbicide–water or herbicide–penetrant mixtures to the outer circumference of freshly cut stumps or the entire top surface of cut stems. Applications are made with a spray bottle, backpack sprayer, wick, or paint brush. Freshly cut stems and stumps of trees, woody vines, shrubs, canes, and bamboo stems can be treated with herbicide mixtures to prevent re-sprouting and to kill roots. It is critical the cut is made as low as possible to the ground, and the stem is treated immediately after the cut is made. Completely wet the outer edge with the herbicide or herbicide mixture. Make sure the solution thoroughly covers the wood next to the bark of the stump. Completely wet the tops of smaller stumps and all cut stems in a clump. Apply a basal spray mixture of herbicide, oil, and penetrant to stumps that have gone untreated for more than 2 hours. Although winter treatments are slightly less effective than growing season applications, the absence of foliage on some cut stems and branches produces some offsetting gains in application efficiency.

Table 10.2 Susceptibility of trees to common forestry herbicides

Herbicide	Susceptible	Moderate	Tolerant	Highly tolerant
Arsenal AC/ Chopper (imazapyr)	Sweetgum Southern red oak Northern red oak White oak Post oak Water oak Chestnut oak Black cherry	Hickory Dogwood Ash Beech Sourwood Blackgum Red maple		Elm
Accord/Roundup/ generics (glyphosate)	Sweetgum Southern red oak Post oak Blackgum Sourwood		White oak Northern red oak Water oak Red maple Black cherry Dogwood Pine Elm Chestnut oak Beech	Ash Hickory
Garlon 3A (trichlopyr)	Sweetgum Southern red oak Northern red oak White oak Post oak Hickory	Dogwood Pine Elm Chestnut oak Sourwood	Blackgum Water oak Red maple Black cherry Ash Beech	
Pathway/Tordon (picloram)	Sweetgum Southern red oak Northern red oak White oak Post oak Water oak Chestnut oak Black cherry	Beech Hickory Sourwood Dogwood Pine Elm Ash Blackgum	Red maple	

(Continued)

Table 10.2 (Continued) Susceptibility of trees to common forestry herbicides

Herbicide	Susceptible	Moderate	Tolerant	Highly tolerant
2,4-D	Southern red oak		Sweetgum	Water oak
	White oak		Northern red oak	Red maple
	Post oak		Black cherry	Ash
	Blackgum		Hickory	Chestnut oak
	Dogwood		Pine	Beech
	Elm		Sourwood	

Directed foliar sprays and wipes

Directed foliar sprays are herbicide–water–adjuvant solutions aimed at target plant foliage to wet all leaves. An adjuvant is a chemical added to the mixture to make the herbicide more effective. A surfactant is the most commonly used adjuvant and helps the herbicide penetrate the waxy outer covering of leaves. Herbicide application by directed foliar spray is one of the most cost-effective methods for treating many types of herbaceous and woody plant species. Foliar sprays can be applied whenever leaves are present but, for woody plant control, are usually most effective from midsummer to late fall. Winter and spring applications are also effective in controlling certain species.

Herbicides labeled for directed foliar sprays for site preparation that are not restricted to conifer reforestation are

- Accord and Roundup (glyphosate)
- Garlon 3A and Garlon 4® (triclopyr)
- Tordon 101 and Tordon K (picloram)
- 2,4-D and Weedone 2,4-DP®

Tank mixes of these products will usually be more effective when treating mixed species.

Basal sprays and wipes

Basal sprays are herbicide–oil–penetrant mixtures sprayed on the lower portion of woody shrub, vine, and tree stems. They are usually applied with a backpack sprayer. Basal sprays are best where most trees are less than 8 inches dbh but can be used on much larger trees of susceptible species. Apply to smooth juvenile bark by thoroughly wetting the lower 10–20 inches of the trunk (up to 36 inches on larger trees) to the groundline, including the root-collar area and any exposed roots. Smaller trees and shrubs are controlled with less coverage. Avoid spray contact with desirable trees or heavy use within their root zones.

The herbicide must be an oil-soluble formulation and mixed with a special basal oil product, penetrating oil, diesel fuel, fuel oil, mineral oil, vegetable oil with a penetrant, or blends of these ingredients. Appropriate oils will be specified on the label. The most commonly used mixture for basal applications includes Garlon 4 at 20%, a penetrant at 10%, and the oil carrier at 70%. A penetrant is a surfactant than allows the mixture to penetrate and spread such as Cide-Kick®. The oil carrier can be diesel, vegetable, or mineral oil. Vegetable oil is effective and environmentally friendly. Pathfinder II® (triclopyr) is sold ready to use with these oils.

The most effective time period for a basal spray and streamline is June through September, while winter treatments are easier when leaves do not block access and spray. Fall, winter, and late spring applications are often not as effective, though the period from February 15 to April 1 has shown acceptable results. After treating with a basal spray, wait at least 6 months before cutting dead trees because herbicide activity within plant roots can continue for an extended period.

Broadcast herbicide applications

Herbicides labeled for broadcast applications prior to planting hardwoods are

- Accord and Roundup
- Garlon 3A and Garlon 4
- Tordon 101 and Tordon K
- Oust® (herbaceous plant control)

The best control will be obtained by using mixtures of these products if there is a mixture of target species on a site. Oust can be mixed with these other herbicides without decreasing their effectiveness to increase herbaceous control. When competition is essentially only one species, then one herbicide may be best. Both aerial and ground sprayers can apply these herbicides. Aerial broadcast applications are commonly used on tracts of 50 acres or more because of improved coverage and costs compared to ground applications.

Utility skid and trailer-mounted sprayers

Complete spray systems are available mounted on utility skids for hauling in truck beds or trailers. Sprayer pumps are usually powered and require availability of adequate fuel and oil on-site. The chief benefit of mounted sprayers is their capacity for holding and applying high volumes of herbicide–water mixtures. But their benefit is checked by such limitations as accessibility to the infestation, hose length, and weight and availability of water in the field. Of mounted sprayers, the skid-steer-mounted sprayer is best for getting into difficult terrain and stands.

All-terrain vehicle and recreational-type vehicle-mounted sprayers

ATV sprayers are best for selective applications in sensitive areas. Newly designed sprayers can hold 16, 24, or 40 gal with optional front tank add-ons for many ATV models. Recreational-type vehicles (RTVs) are larger than ATVs and carry a larger amount of herbicide mixture, with a capacity of up to 150 gal. ATVs, with a tighter turn radius, are suited for repeated back-and-forth narrow swaths.

Tractor-mounted sprayers

Spray systems can be mounted on farm tractors, four-wheel drive tractors, skidders, forwarders, and crawler tractors. Tractor-mounted sprayers are useful for large prairie and forest restorations as well as right-of-way projects. Tractor-mounted tanks have a large capacity (some hold up to 600 gal of spray solution per tank), supporting a much larger workload than other ground equipment. Spray nozzle systems for tractors in forested situations are usually boomless, which means the spray comes from a single nozzle.

Backpack mist blowers

Broadcast applications can be made with a gasoline-powered backpack mist blower. A wind turbine creates fine droplets that penetrate into shrub foliage, but these droplets readily drift with wind and fog. As a safeguard to nontarget plants, foliar-active herbicides are usually recommended. These applications are only suitable for internal lands with dense low stands where drift will not move to nontarget plants or lands. Wind must be minimal and moving away from a lane or spot of application into the target foliage.

Aerial sprayers

Helicopter sprayers can apply herbicides on large or remote sites. With GPS technology, helicopter applicators are extremely precise in treating target areas with preprogrammed swaths. They are highly maneuverable and apply sprays at much slower speeds than fixed-wing aircraft.

Contract aerial applicators are available in every part of the South. The land manager or owner also has responsibilities for preparing the site for aerial treatment, such as felling tall snags, heliport construction, and marking boundary lines visible from the air.

Site preparation on old-field sites

The use of disking and subsoiling should be considered for improving soil and competition conditions before planting oaks on abandoned fields and pastures. Disking treatments will improve planting operations if performed correctly and often promote annual herbaceous plants that are more effectively controlled with herbicides than perennial plants. Disking should be to a depth of 8 inches and should be done in strips along the contour to reduce the chance of soil erosion. Unfortunately, disking can aggravate wet weather planting the following spring. Subsoiling or ripping can be used to breakup plowpans that are common to these sites. Part of the decision to use tillage treatments must consider whether the site is designated as wetlands and whether the tillage practice would be considered sod-busting that might jeopardize participation in USDA-sponsored programs.

For controlling invasive grasses and/or forbs before planting, applications of glyphosate herbicides can be broadcast or sprayed in bands to form planting rows. Only Roundup appears specifically labeled for this situation. Rates of 2–5 quart/acre typically will be required to control established sod and even then complete control cannot be expected. A prescribed burn in early summer before late summer Roundup applications can clear standing dead grass and thus improve herbicide efficiency. Also, mowing can be used to improve access and allow better spray coverage by reducing vegetation to a more uniform height. If you mow before an herbicide treatment, wait for 4 to 6 inches of regrowth in warm-season grasses and 8–10 inches in cool-season grasses before Roundup applications.

Site preparation

The easiest way to solve competition problems is to address them before planting. A site preparation spray will alleviate problems with perennial woody plants that will plague the stand throughout its cycle. By killing woody shrubs and undesirable trees, you give planted trees the opportunity to capture the site without interference from the most rapidly growing vegetation. Herbicides for site preparation can be sprayed with either ground equipment, such as skidders or tractors, or via air by helicopter.

Fixed-wing aircraft are not generally used in spraying herbicides because of their tendency to produce finer particles that may move off site. Nozzles used in helicopters produce raindrop-size droplets with very few fine drops and spray booms for ground equipment are applied low to the ground, which minimizes the movement of herbicide. In choosing herbicides for oak planting, care must be taken in choosing herbicides that do not have soil residual activity that would damage the seedlings. Herbicides such as glyphosate (Accord) and triclopyr (Garlon) do not have soil activity that will affect oaks. Site preparation herbicides, such as imazapyr (Arsenal) and hexazinone (Velpar®), should not be used because of their soil activity. In addition to the herbicide, a surfactant is added to help the herbicide penetrate the waxy cuticle of plants. In fact, most foliar applications will not be effective without the addition of a surfactant. Any nonionic surfactant should be effectual at a rate up to 2%, but waxy leaf competition may present some unique problems. The best time to apply the proper herbicide is from mid-June until leaf drop with the later applications being more effective. Earlier applications may be used if a burn is desired to gain accessibility and allow for more drying time after the vegetation dies to insure a hot fire.

Herbaceous weed control

The majority of oak planting is done on retired agricultural sites. These areas are typically highly productive and heavily colonized by a wide variety of grasses and forbs. Research has demonstrated competition from these plants has a serious impact on the survival and growth of planted oak seedlings. Site preparation, whether mechanical or chemical, will not provide residual competition control needed during the first growing season. Therefore, it is very important that you find a cost-effective method for controlling these plants during the first growing season. While competition control during the second growing season may provide some benefits, research has indicated it is not cost effective.

While a number of mechanical methods are available, none provide an attractive option. Mowing (or bushhogging) will reduce aboveground biomass and can reduce shading, but does little or nothing to reduce root competition. Mowing must also be repeated throughout the growing season and can do nothing to reduce competition in the planted row, which is the exact place oak seedlings need competition control. Thus, the lack of desired control and cost of repeated applications make these options ineffective and undesirable. While many landowners like the appearance of having the areas between rows mowed, it provides little or no benefit to the oak seedlings and destroys early successional habitat used by many wildlife species.

Disking provides short-term control, but again, does not control the vegetation immediately surrounding the oak seedling. The practice is expensive, must be repeated throughout the growing season, and often results in damage to the oak seedlings if the operator tries to get close enough to actually control some of the vegetation impacting the oaks. Disking also stimulates the seed bank and encourages germination of new plants. For all these reasons, disking is not recommended for weed control among planted oaks.

The best approach to controlling undesirable herbaceous vegetation is the use of herbicides. Land managers have two options for applying herbicides for this purpose—either as part of a chemical site preparation application or as a separate application that is completed soon after the oaks are planted.

Chemical site preparation is recommended only when the area is occupied by species that threaten to overwhelm planted oaks, and these undesirable species cannot be controlled using herbicides approved for herbaceous weed control. As noted earlier, the influence of having some nonoak woody competition (perhaps as many nonoaks as oaks per acre) can be beneficial to the growth and development of oak. However, if a chemical site preparation is to be used, Oust XP® (or other sulfometuron methyl products) can be added to the site preparation application, and it will provide residual herbaceous weed control for the first growing season. This practice works very well in establishing pine plantations and has been tested successfully with oak plantings. The addition of 3 oz Oust XP to the site preparation mixture applied during the late summer to early fall will provide very good control (of susceptible species) through July or August of the first growing season. Depending on the species present, control may last through September.

In most situations, the only site preparation needed for planting oaks will be subsoiling. This will provide multiple benefits. First, any restrictive layers will be broken and the oak seedling root development will be facilitated. Second, planting will be enhanced as it is easier to plant in the loosened soil of the subsoil trench. Third, and perhaps most importantly for herbaceous weed control, the subsoil trench provides a well-defined demarcation for a row in which to plant oaks. This allows you to use banded applications (spray a swath over the top of the planted row) as opposed to a broadcast application (spray the entire area) for plantings where no rows can be identified or followed.

On areas where trees are planted following subsoiling (or no site preparation), herbaceous weed control is best achieved by applying the proper herbicide after planting. While more than 50 herbicide products have been tested for use with oak seedlings and many have proved effective, only a few are labeled for use in oak plantings with spraying over the top of the planted seedling. Again, cost-effectiveness is a primary consideration and the following applications are considered the best choices for use with oaks:

- Oust XP® (2–3 oz/sprayed acre)
- Goal 2XL® (64 oz/sprayed acre—preemergence or 32 oz/sprayed acre—postemergence)
- Select® or Fusilade DX® (rate varies by species and growth stage)

Oust XP: Oust XP contains sulfometuron methyl. Do not confuse this product with Oust Extra, which contains both sulfometuron methyl and metsulfuron methyl, the latter of which is not approved for use on oaks. Oust XP controls a very wide spectrum of both grasses and broadleaf weeds. Research has consistently demonstrated 2 oz/sprayed acre provides very good to excellent results, but 3 oz/ sprayed acre may be desirable in some situations depending on the weed species present (consult the label for a listing of species controlled). Oust XP is pH-sensitive and demonstrates more activity at a higher pH. A good rule is to reduce the rate of application by 25%–50% when soil pH is 6.5 and higher. This material is applied in a spray swath of 5- to 6-feet-wide with the oak seedling in the center of the sprayed swath (banded application). Proper timing for the application is post-planting and pre-budbreak. Thus, if the trees are planted in January or February (vast majority of oak plantings), wait 1–2 weeks for the soil disruption from the planting to settle (one rain will do this), and apply the herbicide using 10–15 gal total spray volume per sprayed acre. Note the application should be completed prior to the oaks breaking bud for optimal results, and no surfactant should be used. Also, note the Oust XP label prohibits use of the herbicide in areas of standing water, and you should not use this application if the area is expected to flood after the application. Done properly, this application provides excellent results.

Goal 2XL: Goal 2XL herbicide contains oxyfluorfen and is labeled for both preemergence and restricted postemergence applications over many, but not all species of oaks. This herbicide is widely used for establishing cottonwood plantations and is very effective on a wide variety of forbs, but has little activity on grasses. While this herbicide does not control as many species as Oust XP, it is quite useful in areas where flooding precludes the use of Oust XP. Goal 2XL is also applied as a banded application, but either Goal 2XL or Oust XP may be used in a broadcast application if a planted row cannot be followed. When properly applied, Goal 2XL will bond to the soil and not move with water. Since this material is degraded by heat, it will not provide control as long as Oust

XP in the South. In fact, the same rate of application will last many weeks longer in cooler climates such as Minnesota than in warmer regions such as Mississippi. Research has demonstrated 64 oz/sprayed acre of Goal 2XL is an optimal rate to use as a preemergence application. A nonionic surfactant should be added at 0.25% volume of surfactant/volume of solution if any vegetation is growing on the site at the time of application. These preemergence applications will provide control of broadleaf plants until May or June, and while this is not as long as the control resulting from Oust XP applications, it provides the oak seedlings with reduced competition for the first weeks of the growing season, which is critical to successful establishment. Goal 2XL has a limited (not approved in all states) label for postemergence applications over oaks. The approved rate is 32 oz/sprayed acre and nonionic surfactant should be used at 0.25% volume of surfactant/volume of solution. This postemergence application is not widely used, but it is very effective on many broadleaf species and most species of morning glory if completed before the vines exceed 3–4 feet in length. While generally not as cost-effective as Oust XP applications, the use of Goal 2XL offers a much needed option for many areas where oaks are planted.

Select or Fusilade DX: These products are both grass-selective herbicides and control grass species only and will not control any forb species. While this may seem to be a less desirable control option, there are situations that dictate the need for grass control only. Examples would be a substantial presence of Bermuda grass or Johnsongrass in an area containing numerous species of broadleaves and other grasses that would require the use of Oust XP, an area that floods and has a significant component of grass in the weed complex, or a Bermuda grass pasture or hayfield that did not get a site preparation treatment or that the Bermuda grass did not respond to site preparation. Neither Oust XP nor Goal 2XL will control Bermuda grass or Johnsongrass, and both of these species may be found in retired agricultural areas. These are very aggressive and competitive species, and if they pose a threat to the oaks, they must be controlled for best results.

Unlike Oust XP or Goal 2XL, both of which are most effective when used as preemergence applications, the grass herbicides are postemergence only (foliar active only) and must be applied to actively growing grass to be effective. While both crop oil and nonionic surfactants can be used with these herbicides, the use of a nonionic surfactant (0.25% volume/volume) is recommended to avoid any potential negative effect on the oak foliage that could result from the use of crop oils.

Select contains clethodim and the active ingredient in Fusilade DX is fluazifop-butyl. For optimal results, a split application should be completed when using Fusilade DX. For example, if the proper rate is 20 oz/sprayed acre, 10 oz/acre should be used in the initial application, and

10 oz/acre should be applied 30 days after the initial application. Results of Select may be enhanced by using a split application, but very good to excellent control can be obtained from a single application. Both these herbicides provide excellent results when used properly. These treatments can suppress Bermuda grass but it will not be eliminated. None of these herbicides will control everything, and it is important to match the herbicide to the requirements of the site (undesirable species, flooding, etc.).

Research over the past 25 years has consistently shown when good seedlings are planted properly, one-third will typically be lost to herbaceous competition in years of average rainfall. Controlling this competition generally results in survival of greater than 90% during these years. During drought years, survival in areas without competition control will drop to 40%–50%, while areas with herbaceous weed control will typically average 80% or greater. Years of extreme drought only make the importance of competition control more dramatic. Oak seedlings planted in retired agricultural areas have plenty of light and the competition for soil moisture is extreme. Herbaceous weed control will promote both survival and growth in these areas.

The use of glyphosate (Roundup®) for herbaceous weed control is not recommended. Glyphosate will harm/kill the oak seedlings and although applicators may be able to complete one application and manage to keep all the spray off the planted oaks, field trials have demonstrated repeated application of glyphosate, despite the applicators caution, will damage or kill the oaks.

Final thoughts

Competition control is critical to the establishment of oaks. It can accelerate the development of natural stands and is a requirement for plantations. Planning is essential for success and the wrong time to plan for competition control is when you have seedlings already planted. It takes at least a year to prepare a site to plant oaks and there are no shortcuts. There are a few important points to remember:

1. Perennials are easier to control before planting, and a site preparation spray is essential.
2. Mechanical treatments can be effective but are expensive and generally difficult to apply.
3. Follow the label for all herbicides.
4. Avoid soil-active herbicides that may kill oaks.
5. Herbaceous weeds affect the growth and survival of oaks and must be effectively controlled the first year.
6. Banding herbicides in rows is effective for herbaceous weed control in plantations.

7. Garlon (triclopyr) provides effective control of woody stems in natural stands.

8. Repeated glyphosate applications using directed sprays result in damaged and dead oaks.

9. Success requires collection of proper information, evaluation of options and proper application.

10. The most expensive treatment you can apply is one that does not work.

chapter eleven

Intermediate treatments*

Wayne K. Clatterbuck

Contents

Crop Tree Release 154
Midstory removal 157
Thinning 158
When to thin 159
What to thin 160
Potential for high-grading 162
Further suggested reading 164

"Intermediate treatments" is a term used for any treatment designed to enhance growth, quality, vigor, and composition of a forest stand after establishment or regeneration and prior to final harvest or maturity. Intermediate treatments are conducted in immature stands for the purpose of improving existing stands by regulating growth through adjustments in stand structure (vertical and horizontal space), species composition, and stand density (number of trees). The objective is to shape the future growth of the stand to some desired condition or structure.

This chapter focuses on practices commonly used to enhance oak growth and development. Brief definitions of other intermediate operations are also given.

Unfortunately, many hardwood stands are partially cut as an intermediate operation without consideration of the future, through a practice called *high-grading*. When this happens, the healthiest and highest-quality trees are removed too early, while those of poor health and quality remain in the stand.

Although some intermediate operations will not be discussed in detail, their definitions and primary use as intermediate practices are provided here.

Weeding: A release treatment conducted during the seedling stage to suppress or eliminate primarily herbaceous plants that overtop or interfere with desirable young trees. Weeding is generally necessary

---

* The author of this chapter relied on the available scientific literature, but for ease of reading, we have not cited references in the body of the text. Credits for that information are given in the "Further Suggested Reading" section at the end of this chapter.

when site preparation for new reproduction was ineffective or not implemented.

Cleaning: A release treatment made in the sapling stage to free favored trees from less desirable trees of the same age class, which over-top desired trees or are likely to do so. Cleanings are used when an unwanted species partly covers the sought-after species and will continue to dominate the stand. Unless released, the smaller trees will continue to decline and eventually die.

Liberation cuts: A release treatment made in a stand not past the sapling stage that frees the favored trees from competition of older, overtop-ping trees. Liberation cutting usually corrects a problem not addressed by earlier site preparation in eliminating undesirable residual trees. The release promotes vigor of a suppressed sapling understory.

Improvement cutting: Trees past the sapling stage are cut to improve stand composition and quality by removing less desirable trees of any species. Generally, stand improvement removes damaged, defective, or poorly formed trees to enhance growth and develop-ment of the trees that remain. The condition of defective trees did not result from competition with other trees. Improvement cut-ting removes trees damaged by natural agents such as wind or ice, freeing healthier adjacent trees in the upper canopy for improved growth and development. The practice serves as the first step for improving degraded stands. Thinnings follow in subsequent years to adjust stand densities and manage growth.

Precommercial thinning: A thinning that does not yield trees of commer-cial value, usually designed to reduce stocking in order to concen-trate growth on more desirable trees.

Sanitation cutting: The removal of trees to improve stand health by stop-ping or reducing the spread of insects and disease. Sanitation cuts are applied in anticipation of the advance of some injurious agent to reduce pending financial losses and minimize chances that the agent will spread to uninfected trees.

Salvage cutting: The removal of trees that are dead, damaged, or dying from injurious agents (e.g., ice, insects, fire) to recover value that otherwise would be lost. Salvage cutting is implemented after the trees are damaged.

Stand improvement: A term comprising all intermediate cuttings made to improve the composition, structure, condition, health, and growth of stands.

Crop Tree Release

Crop tree release (CTR) is an intermediate treatment that provides increased growing space to selected trees through removal of crown competition from adjacent trees. Although CTR could be considered a special type of thinning, traditional thinning techniques are intended to reach a desired stand density or remove specific sizes or classes of trees. CTR differs from traditional thinning in that most resources are focused on a small number of selected trees.

Factors that affect stand value are species composition, tree size, and tree quality or form. By releasing crop trees, diameter growth is increased through expanding crowns and greater photosynthesis. Favoring the more valuable or desired species enhances species composition, and stand quality is improved by selecting crop trees with straight stems, without visible stem defects, with the potential to increase in tree or log grade. The stand progressively becomes more valuable by increasing the proportion of selected crop trees and reducing the proportion of low-value competitors.

CTR is practiced for a fairly narrow range of stand conditions when trees are young (approximately 10–30 years), the stand is crowded or overstocked, and faster-growing trees jeopardize the growth of potential crop trees. For example, in a mixed hardwood stand, slower-growing oaks may be preferred to a faster-growing species such as yellow poplar or sweetgum. Before the faster-growing species fully dominates the stand and the growth of oak slows down appreciably from inter-tree competition, CTR is performed to give the selected oaks more space for continued growth and development. CTR should only be applied on the more productive sites where trees will respond readily to release. On the poorer sites where stems do not respond as strongly to release, the costs associated with implementing CTR may exceed returns.

CTR should be applied in young stands when trees are approaching crown closure (overstocked) and have developed straight merchantable stems. Target 30–50 crop trees per acre with the desired features for crop trees. Remove only those trees interfering with crown expansion of crop trees. Use a crown-touching technique to release the crop tree on at least two sides, preferably three or four sides, for lateral crown expansion. Overtopped or suppressed trees are not influencing the lateral spread of the crown and can remain. Competing trees may be cut or girdled, chemically treated or both.

Generally, species is the primary factor in selecting crop trees, even though other tree attributes can also be used. Market value, wildlife value, and more subtle benefits such as visual quality, recreation, and diversity are determined by stand species composition. Choose crop trees that are healthy and a preferred species based on your management objectives, have the potential to grow and remain competitive in the stand, are relatively straight with few forks, are the better grades (with few defects or knots) in the upper levels (usually codominant crown class) of the canopy, and young (10–30 years). Trees in stands that are too young have not expressed many of the criteria to be considered as crop trees, and trees in older stands might not successfully respond to the release. CTR can also be used to maintain species diversity. If a particular species is relatively scarce or has only a few remaining representatives in a stand, it should be included on the CTR prescription to ensure species diversity.

Avoid selecting trees for CTR if they are in the intermediate or suppressed crown classes, have a flat-topped crown, or have sparse, spindly crowns with less than 33% live crown ratio (ratio of live crown length to total tree length). Trees with these characteristics do not possess the vigor needed to respond appreciably to CTR.

CTR also has many wildlife benefits. Dead standing trees create snags that provide food for insects, which become food for birds, as well as sites for nesting, roosting, denning, and perching. Felled trees provide groundcover and coarse woody debris important to many small mammals, reptiles, and salamanders. Dead trees allow sunlight to reach the forest floor, increasing forage, and nesting cover. Increased sunlight also allows the crowns of crop trees to expand, increasing the amount of area for mast production.

Competing trees are controlled by two methods—physically cutting or girdling the tree with a chainsaw or brush saw and by herbicide application. Cutting trees with a saw is suitable when the number of trees to eliminate is relatively low. Felling trees can be difficult, time consuming, and a safety hazard. If cutting is used, no herbicide is required unless there is a need to prevent resprouting of cut stems, especially when treating an undesirable or nonnative species that might be a nuisance in later regeneration phases.

In most stands, competing trees can be controlled using herbicides (see Chapter 10). Cut-surface methods, such as hack-and-squirt or basal tree injection, administer herbicides through an incision in the bark. Basal bark treatments can be used on trees less than 4 inches in diameter. Herbicides provide both top and root kill and should be considered when problematic species need to be eliminated. Several common herbicides, such as glyphosate, triclopyr, and imazypyr, can be used in CTR. Select the appropriate herbicide for the species to be controlled (see Chapter 10). Make sure to follow label instructions for safety considerations and efficient application.

With CTR treatments, there is also an element of risk to consider. Crop trees may exhibit branching that could reduce log grade. Damage to crop trees during the release treatment although uncommon, may degrade trees. Released trees have the potential for damage from wind and ice. The increase in value of crop trees must be balanced by the costs of treatment and the potential risks to determine whether the treatment meets management and financial objectives.

Midstory removal

Midstory removal is an intermediate treatment used to develop the size and vigor of advanced oak regeneration so that oak can successfully compete with faster-growing species such as yellow poplar, black cherry, and red maple after a regeneration harvest. This practice is used primarily to promote regeneration of oaks in mixed species stands for both wildlife and timber purposes. Successful oak regeneration comes from stems present on the site at the time of harvest (see Chapter 6). These stems can be either seedlings, known as advanced reproduction, or sprouts from stumps of harvested trees. Seedlings established after harvest do not grow as fast as those with already established root systems (advanced reproduction or sprouts). These new seedlings generally cannot compete with faster-growing vegetation. Thus, advanced oak regeneration should be established several years prior to harvest and at least 4 feet tall so that oaks can compete with other faster-growing species that regenerate when trees are harvested.

Treatments that allow too much light to the forest floor promote reproduction of more sun-loving species like yellow poplar. Not allowing enough sunlight to reach the forest floor favors the development of shade-tolerant species, such as maples and the American beech. The removal of leaf area close to the ground without opening the canopy increases diffused light levels where oak seedlings can survive and maintain height growth, but not so much light that competing more shade-intolerant species are stimulated to germinate and grow. Creating gaps in the main canopy allows more light for regeneration of many co-occurring species (yellow poplar, sassafras, and sweetgum) that displace the slower-growing advanced oak reproduction. Thus, midstory removal promotes diffused light levels benefiting oaks compared to other species and minimizes canopy openings.

Research has indicated 20% of stand basal area should be removed during midstory removal. The smallest trees (less than 1 inch in diameter) are removed first, progressing to larger diameters until 20% of stand basal area is removed. Most understory and midstory trees are removed allowing sufficient light to penetrate for oak seedling growth.

Midstory and understory trees should be treated with herbicides when trying to increase diffused light to the forest floor. Simply cutting these trees will lead to sprouting and cause a greater shade problem for the growth of advanced oak reproduction. Using herbicides ensures the elimination of or reduction in competing and undesirable species. The choice and application of herbicides are similar to that described earlier for crop tree release (see Chapter 10). Herbicides should not be applied when the sap is rising in the late winter and early spring since translocation of herbicide to the roots is minimal at this time.

Thinning

Tree mortality is a natural, ongoing process in the forest. Young forests with small trees support thousands of individual trees per acre. As the forest matures and individual trees grow larger, many trees are crowded by faster-growing neighbors and die. The forest naturally thins as trees compete for growing space.

Thinning is a forestry technique that mimics this natural process of mortality. By applying cuts to immature stands, material that might otherwise die before maturity can be used and growth can be concentrated on fewer, more desirable stems left in the stand. Thinning reduces crowding through altering the physical space and sunlight received by the remaining trees. By redistributing growth to fewer, more desirable trees on the site, the overall health, vigor, and growth rate of the residual stand are increased. Each site has an inherent production potential. Thinning allows you to capture that potential for desired "crop trees." Thinning maintains or increases growth of residual trees, reducing the time required for trees to reach a prescribed diameter. Thinning can also provide some intermediate return on a landowner's long-term investment (see "Why Thin?").

WHY THIN?

Many forest stands are overcrowded. Without growing space for trees to continue to grow and expand their crowns, trees growing too close together become stressed and unhealthy. Overstocked (too many trees) stands usually have the following characteristics:

- *Poor tree health*: Overcrowded and stressed trees have poor diameter growth and small crowns. Stressed trees are more susceptible to drought and insect attack.
- *Poor wildlife habitat*: If too little sunlight reaches the forest floor, shrubs, grasses, and forbs, which are beneficial to wildlife, are shaded out.
- *Unattractive forest*: Most people find a well-spaced forest more inviting, accessible, and attractive than an overcrowded stand.

Thinning is one of the most beneficial practices to improve the health and vigor of the stand. When branches of adjoining trees touch, it is time to thin. Some guidelines for thinning include:

- *Save the biggest and the best*: Retain the tallest trees with larger diameters and large healthy crowns. These trees are increasing in value at the greatest rate.
- *Remove competitors*: Trees with below-average diameters, shorter crowns, poor form, disease, insects, or other problems should be

removed from the upper-level canopy to benefit their more desirable neighbors. Do not make the mistake of trying to release inferior, lower-level trees by removing larger, better trees growing above them and assuming that these inferior trees will become future crop trees.

- *Retain the best suited species*: Keep the kinds of trees that are most suited for the site.
- *Retain trees for wildlife*: A few dead or hollow trees should be retained unless they are unsafe. Consider leaving a few small clumps of trees to provide habitat diversity. Proven mast producers should be retained.
- *Remove enough trees*: A common error is to remove too few trees, resulting in a thinned stand that is still stressed and overcrowded. After thinning, branches from the crowns of adjoining trees should be several feet apart with open sky visible between trees.

Thinning allows trees to get larger quickly by reallocating the volume growth on fewer trees, thus reducing the time of investment. Thinning should only be practiced with immature stands, to promote more space for the remaining crop trees to grow and become larger.

When to thin

While there is no magic number of years to wait until thinning, several indicators suggest when a thinning should occur. As trees grow and mature, their crowns will begin to compete with each other for available sunlight. Eventually, this crown competition will result in a forest with a closed canopy, where sunlight does not reach the forest floor. A closed canopy indicates tree crowns do not have space to continue to grow and expand. Understory vegetation is sparse because of the lack of sunlight. If these characteristics are present in a stand, a thinning should be applied. A more quantifiable estimate of when to thin based on crown structure is when the live crown ratios are less than 33%. Live crown ratios decrease because sunlight is not reaching the lower branches, decreasing the photosynthetic area of the tree. As a result, the lower branches of the tree die, reducing the overall length of the live crown. Healthy trees maintain live crown ratios of 33%–50%. Another indication of when it is time to thin is when the diameter growth rates begin to decrease, which again, is an indication of less photosynthetic or crown areas.

Several thinnings can be implemented throughout the life of a stand. As trees increase in size in response to a thinning, growing space once again becomes more limited and tree growth declines. Thus, subsequent thinnings should be conducted to maintain tree growth until maturity.

What to thin

Trees to be cut and trees to remain during a thinning operation depend on landowner objectives. For timber, trees of unwanted species, poor form, or slow growth are chosen for removal. With wildlife objectives, consistent mast producers and some with cavities might be retained to provide food or shelter. Also, some species of wildlife prefer the presence of some tree species to others for food, cover, and other needs, and this may influence which trees are retained in a stand. How much to thin also depends on objectives. If it is being conducted as a commercial operation, enough trees must be thinned to make the harvest attractive to a logger. In most thinnings, about one-third to half of the trees is removed.

The focus of a thinning operation should be the trees to retain, not the trees to be harvested. If creating or improving wildlife habitat is an objective, then the resulting structure of the stand may also be a consideration. Thinning operations can create and improve habitat for a variety of wildlife. The increase in sunlight reaching the forest floor after a thinning will promote the growth of forbs and understory shrub species, providing more cover and food for wildlife. Thinning operations can be designed to complement this understory development. For example, some of the trees retained can be grouped into clumps or "islands," instead of individual trees, to create a variety of structure. Trees retained in these clumps will not have the increased space for additional crown growth, so all objectives for the stand and the trade-off of such practices need to be carefully considered by the landowner. Acceptable growing stock refers to trees that are growing or have the potential of growing and increasing in value. These trees should be retained and protected during a thinning. The inferior or unacceptable growing stock should be removed. Refer to the sidebar, "Hardwood Tree Classes," for one method in determining or classifying trees to leave or cut in a thinning operation.

Thinning and associated logging can damage valuable trees that remain after the harvest. Select loggers or timber buyers that will take care of the leave trees (those selected to remain in the stand) during thinning to minimize skidding and felling damage. A damage clause with specific criteria establishing a threshold for tree injuries (basal wounding, bark stripping, branch or stem breakage), and stipulating associated financial penalties, should be part of the timber sale agreement.

Mortality is an inescapable function of nature as stands develop and grow and trees compete with each other for resources. Thinning is an intermediate treatment that optimizes growing space. Thinning speeds up what would happen naturally in the stand and usually favors the better trees in a high position in the canopy. A properly timed thinning will result in bigger, healthier, and faster-growing trees. It can also bring substantial wildlife benefits to a stand through increased canopy structure (gaps plus crowns of various sizes), more "layering" of the midstory through the release of these stems in response to the increased light entering through the canopy, and a flush of understory vegetation.

HARDWOOD TREE CLASSES: A FORM OF CUTTING PRIORITY

Management and Inventory of Southern Hardwoods
USDA Agriculture Handbook 181 (Putnam et al. 1960)

Designating tree classes is an excellent method of determining harvesting priority. Tree classes serve as the basis for choosing trees to cut and trees to leave as well as developing tree-marking rules. The four tree classes are (1) preferred growing stock, (2) reserve growing stock, (3) cutting stock trees, and (4) cull stock trees.

Preferred growing stock (PGS) trees (leave trees) are the future crop trees that will be grown until the end of the rotation. These trees are in good condition, of desirable species, growing at an acceptable rate, in the dominant or codominant crown class, of good grade or with the potential to develop into a high-quality tree, and can be left indefinitely without the risk of dying. These are the best trees in the stand that are increasing in value quickly. Usually, 10%–15% of the trees in a stand are considered PGS trees.

Reserve growing stock trees (storage trees or acceptable growing stock) are in good condition but do not qualify for preferred growing stock. Generally, these trees are not growing as well as PGS and are of poorer grade without the potential to increase in grade. Reserve growing stock can be left for one or more cutting cycles with little risk to merchantability or survival. However, these trees are not the ones left for the final harvest.

Cutting stock trees (cut trees or unacceptable growing stock) are those that must be cut during the next cutting cycle because they are in poor condition and will not survive for future cutting cycles. These trees are usually of an inferior species for the site or of poor form or grade that will not increase in value.

Cull stock trees are incapable of meeting the desired product goals. Most of these trees are taking growing space that would be better suited for the more desirable trees. Cull trees can be recognized as two types: (1) sound cull trees that will never make sawlogs, but contain usable fiber and (2) unsound cull stock trees that do not contain merchantable fiber.

After designating the class of each tree, marking priority for trees to cut or leave is as follows:

- All *cutting stock* trees and *cull stock* trees are cut except in those instances where those trees may be providing visual or wildlife benefits.
- None of the *preferred growing stock* trees are cut except in special circumstances such as a species being universally killed by insects or disease, two preferred trees growing side by side each affecting the growth of the other, or unusual market prices or conditions.
- *Reserve growing stock* trees may be cut or left according to the intensity of the harvest to be made. Often, when only cutting stock and cull stock trees are offered for sale, they are not valuable enough to entice timber buyers. Thus, the sale is "sweetened" somewhat with a few, more valuable reserve growing stock trees to amass enough volume and value to be economically attractive.

The use of tree classes as a cutting priority allows one to provide the best growing conditions to those trees that will continue to increase in value: the fast-growing, desirable species, higher-grade trees.

Potential for high-grading

High-grading (often guised as thinning) is a timber harvest that removes the trees of commercial value only, leaving small trees and larger ones of poor quality or of a low-valued species. High-grading is a purely extractive cutting scheme without consideration of future growth, development, and composition of the forest. The practice is not recommended but commonly perpetuated because of the desire for short-term profit from harvesting high-value species and better grades of hardwood logs, compared to inferior species, lower log grades, and smaller diameter trees.

High-grading has several detrimental effects. First, it removes the better-quality trees, leaving growing space for poorly formed trees or species of low value. Second, high-grading leaves the midstory, which limits the amount of sunlight reaching the forest floor, resulting in long-term development of regeneration from shade-tolerant species such as maples, elms, and the American beech, instead of those species such as oaks, ashes, walnut, cherry, and yellow poplar that require full or moderate sunlight to regenerate. A third problem is in the perpetuation of small, but old trees. Often these smaller trees are just as old as the overstory trees and cannot be relied upon to become future overstory trees.

Too often small trees are equated with young age classes. Although young trees are usually smaller, it is not always true that small trees are young. Different species grow at different rates, differentiating into crown classes: dominant, codominant, intermediate, and suppressed/ overtopped. The most rapidly growing species soon dominate the upper canopy, relegating the slower-growing species to the mid-canopy and understory. What appears to be a cohort of young, small trees ready to occupy the upper canopy when the harvest of the upper canopy is completed may be a stratified mixture of species of a similar age. The trees in the understory and mid-canopy are the same age as the stems in the upper canopy. Instead of being young, vigorous understory trees waiting for the opportunity to grow, they are old, small-diameter stems with sparse crowns and little growth potential. Leaving these trees for the future reduces the volume and quality yield of the stand. A qualified forester can identify young and older trees in a stand even if they are similar in size and provide advice regarding the appropriate trees to harvest to avoid high-grading.

Oak-dominated stands often have understories of red maple, sourwood, blackgum, American beech, redbud, flowering dogwood, and a number of shrubs, but it is rare that the more valued species of oaks, ashes, and others form an understory that will respond favorably to release. Intermediate practices, such as midstory removal, should be implemented to favor the regeneration and development of the desired species before a harvest takes place.

Finally, repeated entries at 10- to 20-year intervals typically associated with high-grading adversely affect the development of good stem form and straight trees as the crowns of these inferior stems tend to expand laterally. High-graded forests are degraded through promoting undesirable species, trees of poor form, reduction of a valuable overstory, and loss of desirable regeneration.

The solution is to avoid high-grading and prescribe thinnings or partial cuts with the objective of perpetuating the growth and development of the remaining trees in the stand. Retain enough trees (acceptable growing stock) so that they use the vacated space of the cut trees to expand their crowns and grow larger. Thinnings will emphasize the trees that will remain and constitute the future forest rather than the trees harvested. If there are not enough acceptable growing stock trees remaining with the potential for a future harvest, then provisions should be made to regenerate the stand for the favored species.

Cutting the largest and most valuable trees is a dysgenic practice that removes trees that are best suited to the site and usually growing at an acceptable rate. The trees that are less adapted remain, usually growing at a low and less acceptable rate of return. The purpose of thinning is to allow the more valuable trees more room to grow and increase in value, not an opportunity to high-grade. The use of tree classes as a cutting priority allows for the best growing conditions for those trees that will continue to increase in value—the fast-growing, desirable species, and better grade trees. Thinning, performed correctly, yields bigger, better, healthier trees with more value at a faster rate and creates an excellent opportunity to improve wildlife habitat.

Further suggested reading

Catanzaro, P. and A. D'Amato. 2006. High grade harvesting: Understand the impacts, know your options. Amherst, MA: University of Massachusetts Extension, 20 pp., http://masswoods.net/sites/masswoods.net/files/pdf-doc-ppt/High_Grade_Harvesting.pdf, accessed August 25, 2015.

Clatterbuck, W.K. 2004. Big trees, little trees—Is there always a correlation with age? *Forest Landowner* 63(1): 26–27.

Clatterbuck, W.K. 2006. *Treatments for Improving Degraded Hardwood Stands*, Professional Hardwood Notes, Publication SP 680. Knoxville, TN: University of Tennessee Extension, 12 pp., http://utextension.tennessee.edu/publications/Documents/SP680.pdf, accessed June 20, 2014.

Meadows, J.S. and D.A. Skojac, Jr. 2008. A new tree classification system for southern hardwoods. *Southern Journal of Applied Forestry* 32(2): 69–79.

Miller, G.W., J.W. Stringer, and D.C. Mercker. 2007. Technical guide to crop tree release in hardwood forests, Professional Hardwood Notes, Publication PB 1774. Knoxville, TN: University of Tennessee Extension, 24 pp., http://utextension.tennessee.edu/publications/Documents/PB1774.pdf, accessed June 20, 2014.

Oliver, C. 2008. Commercial thinning 101: When and what to consider. *Tree Farmer* 27(5): 6–11.

Putnam, J.A., G.M. Furnival, and J.S. McKnight. 1960. Management and inventory of southern hardwoods. Agriculture Handbook No. 181. Washington, DC: U.S. Department of Agriculture, Forest Service. 108 p.

Stringer, J. 2006. Oak shelterwood: A technique to improve oak regeneration, Professional Hardwood Notes, Publication SP 676. Knoxville, TN: University of Tennessee Extension, 8 pp., http://utextension.tennessee.edu/publications/Documents/SP676.pdf, accessed June 20, 2014.

section three

Managing oaks: How do I make it work for me?

chapter twelve

Where am I?

Tamara Walkingstick, Kyle Cunningham, and Jon Barry

Contents

Tree management categories 175
    Preferred and reserve growing stock 176
    Cull stock 176
    Cutting stock 176
Stocking level 176
Woods walk 178
Information on managing oaks 179
State and federal agencies 179
    State forestry agencies 179
    State wildlife agencies 179
    Cooperative Extension Service 179
    Natural Resources Conservation Service 180
    University forestry departments 180
    Consulting foresters and wildlife biologists 180
Forestry management plan 180
Further suggested reading 181

There are many questions that arise when you make management decisions about a hardwood stand.

- What are the management goals?
- What tree species are best for the site?
- What types of timber products can the stand produce?
- For which wildlife species can the stand provide a habitat?
- Is the stand ready to regenerate, or is timber stand improvement the best option?

Managing hardwood forests to promote oaks is not easy. Most hardwood forests contain many tree species with varying growth rates and requirements (Figure 12.1). The first step in developing a management plan for a particular forest stand is to evaluate the current stand and site conditions. This is important for several reasons, including (1) the current

Figure 12.1 A mature hardwood forest. (Photo courtesy of K. Cunningham.)

management potential of a hardwood stand for timber products, wildlife habitat, etc.; (2) species that will grow well on a particular site; and (3) the achievable management goals (Figure 12.2).

The next thing you should consider when managing a forest is whether you want to produce revenue to finance management operations. If so, this consideration places emphasis on timber as an important component of your management plan. The level of emphasis on timber varies greatly with your goals. If timber is the primary goal, then management will be based largely on producing high-quality trees in a relatively short period of time. However, if improving your forest for wildlife is an objective, management will be guided more by providing food and cover for the wildlife you are interested in, and revenue may not even be a consideration. Both goals require stand evaluations, but the variables for evaluation will differ.

Evaluating the management potential of a hardwood stand involves making a decision to either manage the existing stand or regenerate it (Figure 12.3). This decision should be based on whether there are enough desirable trees present for a given objective to continue to manage it (see Chapters 14 and 16). Intensively managed hardwood stands are generally evaluated on 5–10-year intervals, called *cutting cycles*. Therefore, a decision to manage an existing stand means (1) the existing stand will be managed another 5–10 years and reevaluated and (2) the direction of stand management will involve intermediate stand management operations

Figure 12.2 Conducting a timber stand evaluation is the first step in the decision-making process. (Photo courtesy of K. Cunningham.)

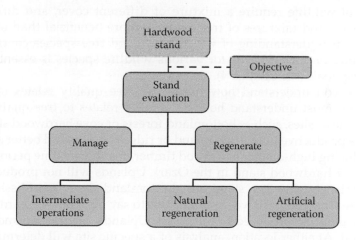

Figure 12.3 Basic flowchart in stand decision making. (Adapted from Cunningham, K.K. et al. 2011, *South. J. Appl. For.*, 35(4), 184.)

(see Chapters 9 through 11). During the cycle, intermediate stand management operations could include crop tree release, stand improvement cuttings, and other operations. If a hardwood stand is ready to regenerate (adequate advance oak regeneration present) and this fits your management goals, then management operations should focus on establishing a new forest.

Hardwood stands with low management potential are easy to identify because there are few desirable trees or they are inaccessible because
of factors such as steep slopes or other physical limitations. Hardwood
stands with high management potential are easy to identify because they
generally contain many high-quality trees. Stands with a management
potential in between these extremes create uncertainty for many forest
landowners. The assistance of a forester with hardwood management
experience can be invaluable when evaluating a stand.

Setting achievable management objectives is important (see
Chapter 13). Hardwood stands are unique in that many objectives can be
achieved simultaneously. Goals may include timber production, wildlife,
aesthetics, water quality, recreation, or others. When timber production is
the primary objective, understanding which hardwood grades produce
certain products and applying that to the forest's ability to produce those
products are essential first steps in stand management. Similar considerations must be made for other management objectives. For example,
if improving the stand for wildlife is the primary goal, then you must
understand if and how the trees in the stand are benefiting wildlife, and
what the cover needs are for the wildlife you are interested in. Many
species of wildlife require a mixture of different cover, and different
tree species and mixtures of tree ages are more beneficial than others.
Gaining an understanding of which hardwood tree species contribute
to the food or shelter needs for various wildlife species is essential to
achieving goals (see Chapter 4).

Once you understand how hardwood tree quality relates to tree
value, you must understand how site quality relates to tree quality. In
general, mesic sites, such as bottomland forests or cove hardwood stands,
are more productive than drier upland or ridge-top sites and better suited
for producing high-quality hardwood timber in a shorter time period. For
example, a hardwood stand in the Ozark Uplands will not produce the
same timber quality as a stand on a bottomland site in the Mississippi
Delta (see Chapter 3). However, that is not to say high-quality hardwood
timber cannot be produced in the Ozark Uplands, or drier upland sites
in general. At either location, analysis of a specific site will determine its
potential for producing high-quality hardwood timber. Although both
sites may be capable of producing good hardwood timber, the Delta
will produce it in much less time (50–60 years versus 70–80 years in the
Ozarks). Furthermore, the desired tree size (diameter at breast height or
DBH) would be larger in the Delta as compared to the Ozarks (for example, 20- versus 16-inch DBH). Similar comparisons can be made between
forests in the Appalachian Plateaus, Piedmont River valleys, and other
Eastern U.S. regions. Chapter 14 provides further discussion specifically
for managing upland oaks, and Chapter 16 discusses management of bottomland oak forests.

Foresters commonly evaluate the diameter of a tree at a standard height aboveground to ensure consistency. Because trees have more swelling as you get closer to the ground, a measurement there would result in a greater diameter than if measured 2 or 4 or 6 feet aboveground. Years ago, the standard was developed at 4.5 feet aboveground. At this height, most of the swell is gone and it is a convenient height for most adults. In fact, because 4.5 feet aboveground approximates "breast height," the measure has come to be known as "diameter at breast height" or "DBH."

Not only is site quality important, but most hardwood tree species also exhibit specific site requirements. The combination of site and species requirements, referred to by forest professionals as "species-site relationships," is an important factor to understand in hardwood timber management (Figures 12.4 through 12.6). This concept becomes increasingly important when attempting to establish hardwood stands through artificial regeneration. For example, on many bottomland sites, a single hardwood species, such as cherrybark oak, is selected for its desirable growth rate and timber quality. However, cherrybark oak has specific requirements that do not fit every site. Therefore, mixed results may occur with regard to stand establishment

Figure 12.4 Northern red oak on a productive upland site. Such high quality stems can be grown on very good sites. (Photo courtesy of K. Cunningham.)

Figure 12.5 Natural regeneration harvest in a white oak stand on a dry upland site. Note the good quality stems near the drain and the lower quality stems on the higher slope positions where site quality is lower. (Photo courtesy of K. Cunningham.)

Figure 12.6 A cherrybark oak plantation on a minor stream bottom. This extremely valuable species is sensitive to site and will not perform well on uplands or where drainage is poor. (Photo courtesy of K. Cunningham.)

and growth. In many cases, it may be desirable to manage for multiple hardwood species that meet specific site requirements (see Chapter 8).

From a timber production standpoint, among the most important factors in hardwood stand evaluation is the quality of stems present. The quality of hardwood trees is evaluated by the butt log grade. The butt log on a hardwood tree is the portion of the trunk located within the first 17½ feet from the ground. The butt log of hardwood stems generally contains the majority of the value of the tree.

Properly grading the butt log of hardwood trees requires appreciable hardwood management knowledge. The butt logs of hardwoods are typically given one of three field grades—F1, F2, or F3—with F1 representing the highest grade. Factors that determine log grade include size (diameter), amount of clear wood, and straightness. The size of a tree is essential in log grading, because larger diameter trees require shorter lengths of clear wood than smaller diameters to maintain a high grade.

If wildlife is your primary objective, how well your forest meets the food and cover needs of the wildlife you are interested in having on your land is the most important factor. The food and cover requirements of many wildlife species change throughout the year. It is important to consider whether it is possible to provide all of these requirements on your land, and also consider what is available in the forests surrounding your land.

The stand age is another key consideration when evaluating a stand. Typically, hardwood stands managed for timber production are based on 80-year rotations or less. After a hardwood stand surpasses 80 years of age, the trees within the stand are highly marketable and more susceptible to lost vigor and degradation. A loss in vigor and grade can become costly in older hardwood stands (Figure 12.7). If managing for wildlife, longer rotations may be desirable to achieve the desired habitat characteristics that meet your wildlife objectives.

You can determine the age of the stand by knowing the previous management, counting the rings of felled trees (Figure 12.8), or by using an increment borer to take a core from standing live trees. Most hardwood stands are even-aged, meaning the trees are very close to being the same age. Therefore, you do not need to age every tree in a stand; just a handful of trees should provide a good idea of stand age.

The vigor of individual trees within a stand influences their ability to sustain growth and value throughout a rotation. High-vigor trees sustain faster growth rates, are better able to defend against insects and diseases, are less likely to experience degrade and decay, and will generally experience less mortality (Figure 12.9). The recent rate of growth can be determined by using an increment borer to count the number of rings in the outer 2 inches of the stem. Trees that are still growing rapidly (e.g., less than four rings per inch) would normally be considered vigorous. By contrast, those growing slowly (e.g., more than 15 rings per inch) would

Figure 12.7 Internal decay damage in a hardwood tree. (Photo courtesy of K. Cunningham.)

Figure 12.8 Many hardwood species can be aged by counting annual growth rings. (Photo courtesy of K. Cunningham.)

Figure 12.9 Injury to hardwood trees often results in loss of vigor, increased decay, and degradation. (Photo courtesy of K. Cunningham.)

normally not be considered vigorous. Other factors should be evaluated as well. Two other key elements are considered in making a vigor rating based on physical attributes: tree crown and tree bole. High-vigor tree crowns should be full (not one sided) and show little sign of limb dieback or decay. High-vigor tree boles (trunks) should maintain tight bark and show little or no evidence of damage or decay. It is important to remember tree species will differ in crown size, crown depth, and bark characteristics. Therefore, some knowledge about the characteristics of different tree species can be useful when determining tree vigor. It should be reiterated that these characteristics may not completely relate to the physiological health of a tree but are the most practical for field observations.

Tree management categories

The tree characteristics discussed to evaluate the hardwood stand—species desirability, grade, stand age, and vigor—can be used to categorize trees within the stand. These categories include manageable trees (preferred and reserve growing stock), cull trees (cull stock), and undesirable trees (cutting stock). Refer to the sidebar, "Hardwood Tree Classes," in Chapter 11 for more information on what comprises each of these stocking categories. The focus of a stand evaluation should be to determine the number of trees per acre within each category. Categorizing trees into one of these categories can determine whether a stand is manageable or in need of regeneration.

Preferred and reserve growing stock

Simply stated, these are trees of the proper condition to meet the management objectives for a stand. For timber production, these trees should be desirable species, medium to high vigor, proper grade, and manageable age to achieve desired timber products in the desired amount of time. For goals other than timber production (or different levels of management intensity), these ratings may differ. For example, from a wildlife perspective, species, size, and tree age would be more important than grade in determining which trees are manageable. Under an environmental objective, such as water quality, manageable trees may be high vigor, young trees actively absorbing excess soil nutrients. Under this objective, species and grade would have little importance.

Cull stock

Cull trees are a desirable species, but they do not contain the proper grade to harvest for timber products. However, they may be valuable for wildlife. Cull trees may provide cavities for squirrels, bears, or cavity-nesting birds and they may provide mast.

Cutting Stock

These undesirable trees are species not capable of assisting in achieving the desired management objectives. Management operations should focus on removing them.

Stocking level

There is no universal number of manageable trees per acre (TPA) required to consider a stand manageable. The minimum number of manageable TPA depends on the management objective and the size (in diameter) of trees within a stand. A stand that contains tree diameters in the 8- to 12-inch range will require more manageable TPA than a stand with 16- to 20-inch diameter trees.

There is a widely accepted rule that suggests a stand of small sawtimber trees—12–14 inches in diameter—needs approximately 55 manageable TPA. Stocking tables can be useful in making stand management decisions. They provide an estimate of the TPA required for stands of different sizes (average diameter) at different stocking levels. Although these tables are based on the number and size of all trees present, knowing the number of manageable TPA can make these tables more beneficial in the decision-making process.

Figure 12.10 illustrates the number of TPA required for upland hardwood stands of different sizes to be understocked, fully stocked, or overstocked. From a management standpoint, understocked stands contain the minimum number of trees per acre to be considered adequately stocked for management. Fully stocked stands are those with an optimal number of TPA for management. Overstocked stands are those with too many TPA.

If the stand evaluation suggests you should manage an existing stand, then plan for management operations designed to improve the existing

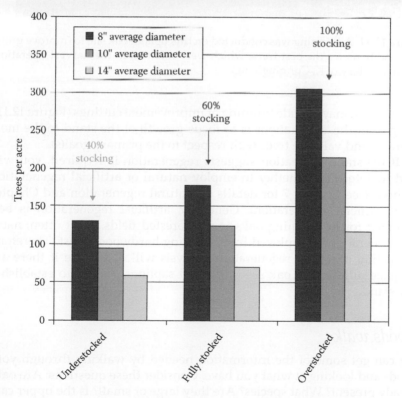

Figure 12.10 A stocking guide for upland hardwood stands. (Adapted from Gingrich, S.F. et al., *For. Sci.*, 13, 38, 1967.)

Figure 12.11 A thinning was conducted on this upland oak site to improve grow-ing conditions for the best trees, which were retained in the harvest operation. (Photo courtesy of K. Cunningham.)

stand, which may include thinning or improvement cuttings (Figure 12.11). The goal of these operations is to focus growth of the stand to the most desirable and valuable trees with respect to the primary goal(s).

If the stand evaluation suggests regeneration is required, you will need to determine whether to employ natural or artificial regeneration methods (see Chapter 7 for details on natural regeneration and Chapter 8 for artificial regeneration). Generally, artificial regeneration is bet-ter suited for establishing oaks in nonforested fields. Most often, natu-ral regeneration is employed in an existing hardwood forest. Therefore, conducting a natural regeneration analysis will determine if there are adequate numbers of oak seedlings and saplings present to establish a new stand.

Woods walk

You can get some of the information needed by walking through your woods and looking at what you have. Consider these questions: Are oaks already present? What species? Are they large or small? Is the upper can-opy open, creating a well-lit understory, or is the upper canopy dense, creating a dark understory? Do the majority of the trees look healthy, or is there evidence of stress or damage, such as numerous dead or broken branches in the crowns, visible decay in the trunks, or trees with poor form?

Information on managing oaks

Oak stands are complicated to manage. You must consider many factors when choosing the proper plan of action for managing your oak forest. In many instances, more knowledge of hardwood management principles is required than you may have available. Fortunately, help is available.

State and federal agencies

State forestry agencies

Most states have an agency that provides technical forestry assistance to landowners. It may be called a forestry commission, forestry service, or department of natural resources. Most agencies have county offices that you can visit to discuss management needs. Price reports, forestry consultant lists, sample management plans, seedling orders, and other important information usually can be found on your state's forestry agency website. State foresters and your state forestry agency are typically one of your best first resources for finding information about managing your forest, and any forest management plan developed for your property should be done in consultation with a state forester. Often, state foresters will work with individual landowners to develop a forest management plan based on your management objectives.

State wildlife agencies

All states have an agency charged with managing wildlife resources. In the Eastern United States, these agencies typically have foresters on staff because these agencies own and manage forestlands. Many of these agencies can send a biologist who specializes in assisting private landowners with management decisions, as well as landowner assistance programs. If one of the goals is to manage forests for wildlife, these are the individuals to call. The agency's website is a good place for more information.

Cooperative Extension Service

In most states, the extension service has either county or regional personnel as well as faculty at the university who can assist you. A forestry agent may be able to visit the property and can help with forestry decisions. There is also useful information on forest and wildlife management on various state extension websites. Videos, fact sheets, and other materials are available, as well as information regarding workshops, field days, and short courses related to forest management.

Natural Resources Conservation Service

The Natural Resources Conservation Service (NRCS) is charged with providing assistance to landowners with the goal of protecting natural resources, especially water quality. NRCS has state conservation officers in every Eastern state as well as agents that work with a subset of counties. This federal agency can also provide information about financial assistance for forest management. Eligibility depends on management goals and the availability of specific programs.

University forestry departments

Most states have at least one university with a school of forestry or a forestry department. The faculty may be able to discuss forestry questions over the phone, but most will not be able to visit the property. Many of the university departments maintain websites with information regarding management needs, links to other agencies, and other useful information.

Consulting foresters and wildlife biologists

A consulting forester or wildlife biologist is typically an independent contractor or may work for a private forestry or wildlife consulting firm. In the Eastern states, most consulting foresters are trained to manage oaks and will understand and use that information to help you achieve your forest management objectives, and wildlife biologists can provide guidance in managing your forest to meet wildlife objectives. In some states, after an initial consultation, the state forester will recommend a consulting forester best suited to your objectives. The consulting forester then will work with you to develop your forest management plan. It is very important to consult with a Certified Wildlife Biologist® if improving your forest for wildlife is one of your objectives. It is also important to have a state forester review any forest management plan prior to actually implementing any forest management.

Forestry management plan

Once specific goals are set for the property, you should work with a forester and, according to your objectives, a wildlife biologist to prepare a written plan for getting the forest from its current state to the desired condition (see Chapter 18).

Good forest management may require some investment. Expenses may include hiring a consulting forester, developing firebreaks, periodic prescribed burning, herbicide applications, and reforestation or afforestation costs if starting with nonforested land. Although there are a variety of financial assistance programs available from state, federal, and private conservation organizations to assist with forest management, the best source of funding is income from periodic timber harvests. Talk to a forester about what activities will be necessary to manage your forest and what those activities will cost.

Like any other investment, the goal is to get a return that is larger than the investment, but how to measure that return will depend on your goals. The return may be measured in dollars of income generated from timber sales or hunting leases, or it may be measured in the satisfaction from producing good wildlife habitat or a place to hunt. Talk to a forester and wildlife biologist about the balance of cost versus income. Is that balance acceptable? If not, there is some room for adjustment. A forester can help find a level of investment that will meet specific resource needs. A forester and wildlife biologist should work with you to create the forest management plan that integrates all these considerations and help guide the management of your forest.

Further suggested reading

Belli, K.L., Hart, C.P., Hodges, J.D., and J.A. Stanturf. 1999. Assessment of the regeneration potential of red oaks and ash on minor bottoms of Mississippi. *South. J. Appl. For.* 23(3):133–138.

Cunningham, K.K., Ezell, A.W., Belli, K.L., Hodges, J.D., and E.B. Schultz. 2011. A decision-making model for managing or regeneration southern upland hardwood forests. *South. J. Appl. For.* 35(4):184–192.

Gingrich, S.F. 1967. Measuring and evaluating stocking and stand density in upland hardwood forests in the central states. *For. Sci.* 13:38–52.

Loftis, D.L. 1990. Predicting post-harvest performance of advanced red oak reproduction in the Southern Appalachians. *For. Sci.* 36(4):908–916.

Manuel, T.M., Belli, K.L., Hodges, J.D., and R.L. Johnson. 1993. A decision-making model to manage or regenerate southern bottomland hardwood stands. *South. J. App. For.* 17(2):75–79.

Putnam, J.A., Furnival, G.M., and J.S. McKnight. 1960. *Management and Inventory of Southern Hardwoods, Agriculture Handbook* No. 181. Washington, DC: USDA Forest Service, 102 pp.

Sander, I.L., Johnson, P.S., and R.F. Watt. 1976. *A Guide for Evaluating the Adequacy of Oak Advanced Reproduction,* General Technical Report NC 23. St. Paul, MN: USDA Forest Service, North Central Forest Experiment Station, 7 pp.

chapter thirteen

What are my management objectives?

Daniel Yaussy, Patrick D. Keyser, and Bryan Burhans

Contents

Multiple objectives 183
Timber objectives 184
Wildlife objectives 187
Ecosystem restoration 188
Other objectives 189
Landbase considerations 189
Conclusion 190
Further suggested reading 191

You may be reading this book because you want to manage your property to maintain oak-dominated forests or woodlands that provide high-quality wildlife habitat. But what species or combination of species of wildlife would you like to see on your property? Do you want to sell some of your timber to provide some income? Is timber production your main objective with wildlife habitat an added bonus? How can you mesh the two objectives? Is your property adequate in both size and quality to meet your objectives?

Answer these questions before making any decisions about which management practices to use on your property. The material in this chapter will not answer these questions for you, but it will provide some information and context that will make arriving at the most appropriate objectives for your property easier and, we hope, more satisfying.

Multiple objectives

You, like most forest landowners, likely have several objectives on which you will base your management decisions, which may include wildlife habitat, timber production, aesthetics, long-term investment, reducing tax burden, ecological restoration, carbon credits, biofuel production, or providing a legacy for future generations. You can achieve many of these objectives at the same time with proper planning and an understanding of specific trade-offs certain objectives require. For example, harvesting timber can improve wildlife habitat for certain species of wildlife and also result in a positive cash flow for the landowner.

However, the extent to which you place emphasis on a specific objective impacts your ability to achieve other goals for your land. If your overall objective is to manage for maximum timber production, you also may produce a forest structure that limits habitat quality for some wildlife species. A careful and balanced ranking of objectives is necessary to determine the desired conditions for your property and the steps needed to realize those conditions.

Timber objectives

Historically, prices for standing trees (stumpage) indicate you must produce sawtimber—usually trees larger than 12 inches DBH (see the sidebar on "DBH" in Chapter 12)—to make a profit. Stumpage prices for pulpwood are usually too low for pulpwood to be a profitable management objective by itself (see Chapter 12). Some sites are poor enough that sawtimber-sized stems cannot be produced in a reasonable time frame (less than 60 years), or even at all. Other sites may be able to grow sawtimber, but the quality (and therefore, value) may be low.

The value of each species of hardwood sawtimber is determined by its market demand as furniture, cabinetry, and flooring. There is no formula to determine which hardwood species will have the highest market share in coming decades. The volatility of the market for hardwoods is illustrated by comparing what species are shown in the furniture markets of High Point, North Carolina, and major furniture trade shows. The proportion of oak dining room sets shown increased from the early 1960s through 1990 but has decreased ever since, as maple and cherry have become more popular (Figure 13.1). Kitchen cabinets on display at trade shows went from 55% oak and 5% maple in 1989 to 10% oak and 40% maple in 2004. Therefore, decisions about what species mix to manage for today may have little effect on the profitability of future timber sales. However, oak has remained a valued species through many market cycles of the past century or more, regardless of increased demand and prices for maple, birch, poplar, and walnut. It is not as difficult to produce a future stand of maple sawtimber as it is the one dominated by oak.

Making oak timber production a high priority will require careful attention to managing the density of your forest stands. Competition plays a big role in stressing trees, and stressed trees are more likely to become infected by insects and diseases. If a stand is more than 100% stocked (Figure 13.2), the trees are highly stressed, resulting in significant mortality.

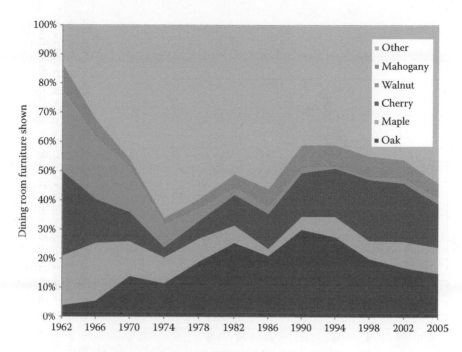

Figure 13.1 Percentage of dining room showings featuring major hardwood species at the High Point, North Carolina, furniture market, 1962–2005. (Adapted from Luppold, W.G. and Bumgardner, M.S., *Wood Fiber Sci.*, 39(3), 404–13, 2007, Table 2.)

Sawtimber stands grow best and are healthiest when they are kept somewhere between 60% and 80% stocked. In a young sawtimber stand with trees averaging 12–14 inches DBH, this healthy range has 50–70 trees per acre, a density that provides for larger crown sizes, improved acorn production, faster growth, and better structure for some wildlife. Reducing the number of trees per acre beyond these levels tends to increase the number of epicormic branches on oaks. These small branches develop on the trunk and reduce timber quality and value. Understocking also reduces total stand yield. Proper stocking may be maintained by the management practices outlined in this book (see, especially, Chapter 11).

There are several logistical and financial considerations that should be evaluated as a part of setting a timber management objective for your property. First, is your property large enough to make various silvicutural operations profitable? Profitable, sustainable oak timber production can be accomplished on as little as 40 acres—or less. The real key is having enough volume to be attractive to a logger to bid on the sale without having to compromise stand condition (i.e., selling trees that you should allow to continue to grow) when you harvest. Another very important logistical issue is access. Poor access can limit the value and marketability of your timber.

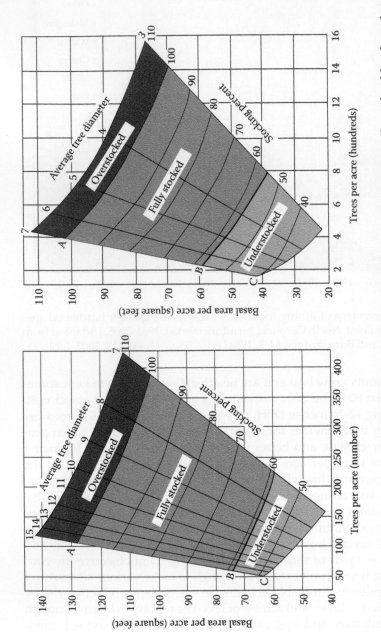

Figure 13.2 The relation of basal area, number of trees, and average diameter to stocking percent for upland hardwood stands. Trees with average diameter 7–15 inches (left); 3–7 inches (right). Stands in the area labeled "overstocked" will experience mortality caused by competition for resources The area between curves A and B indicates the range of stocking where the trees can fully use the growing space. Curve C shows the lower limit of stocking necessary to reach the B level in 10 years on average sites. (From Gingrich, S.F., *For. Sci.* 13(1), 38, 1964.)

Good roads, especially all-weather roads (i.e., improved gravel roads), are essential for implementing any harvest and can increase the value of your timber. If roads do not already exist on your property, they can be expensive to build. However, it is not uncommon to include road construction in a timber sale contract as an offset against revenue due the landowner. Depending on the value of timber to be harvested and the length and cost of road to be built, this may or may not be a viable option.

Periodic income can be achieved through commercial thinnings, shelterwood, and final harvests. Income derived from timber sales can offset other management expenses, such as costs associated with prescribed fire, herbicide treatments, or wildlife habitat improvement. You also have to pay property taxes. Many states have programs to reduce property taxes to offset management costs for land dedicated to timber production. Contact your state forestry agency or extension forest specialist to further explore this option.

Wildlife objectives

Many wildlife species require a variety of forest characteristics and the diverse structure created by multiple timber stand ages to prosper (see also Chapter 4). Migratory songbird fledglings of species associated with mature forests (e.g., wood thrush, cerulean warbler, ovenbird) use berry-producing shrubs found in young forests to forage for several weeks before their first flight south. Studies have found that half the birds found in berry patches within recently clearcut stands were first-year fledglings of interior-forest bird species, feeding on soft mast not available in closed-canopy forests. Ruffed grouse and wild turkeys use herbaceous cover to raise their broods because they are rich in insects, a critical source of protein for fast-growing chicks. They also require overhead cover for roosting, mast production, and protection from raptors.

Managing oak forests for wildlife habitat requires you to decide which species are your highest priority. Any management activity— or lack of activity—alters the forest structure and species composition. Natural succession slowly but continuously changes every forest stand. And each management activity impacts the abundance of each species on your property.

Many game species, and a large number of other of wildlife species, respond favorably to disturbance. So, the use of timber harvest or prescribed fire is essential to achieve certain wildlife objectives. Also, the context in which these disturbances occur (i.e., what habitat is on your neighbors' property) can have a strong influence on how animals will respond to your management activities. Work with a forestry and wildlife professional to formulate a management plan that incorporates your objectives before implementing any such activities.

Keep in mind the more you are interested in wildlife species that use or require young forest cover, such as deer and grouse, the more cutting or burning you may have to do. If you are more interested in species that require mature forests, such as squirrels, salamanders, and some songbirds, less cutting or burning is necessary to sustain these populations. Refer to Chapter 4 and the following chapters for details on these relationships.

Ecosystem restoration

Recent evidence has convinced many forest ecologists that today's oak-dominated forests developed as open woodlands—thanks to frequent, low-intensity, human-caused fire (see Chapter 2). It is likely that most uplands in the oak-dominated parts of the Eastern United States were burned every 5–15 years over the last 3000–5000 years—until as recently as the late 1800s in many cases. Fire controlled oak competition and maintained open-canopy woodlands and ridge-top pines that have been disappearing for much of the past century and have become quite rare today.

As a result of these large-scale changes in oak forests, several federal, state, and local government agencies have chosen ecosystem restoration as the primary objective for portions of the land they manage. Restoring oak woodlands and savannahs involves reducing the number of trees per acre and reintroducing frequent, low-intensity prescribed fires. These practices create a more open tree canopy and promote a diverse ground-cover of plants favored by many wildlife and plant communities. In extreme cases, such conditions become what are referred to as *savannahs* (see Chapters 9 and 15).

There are several ways to reduce trees to the preferred density. However, using prescribed fire combined with a partial harvest is the most efficient and effective way to obtain the desired structure for many wildlife species.

This objective is purely one dedicated to conservation. With the exception of an initial harvest, there will be little or no timber revenue. However, there are trade-offs that could allow for some future cash flow, depending on your objectives. For example, if you are trying to maximize high-quality browse for deer and still maintain mast production, you will want to reduce basal area below levels dictated by timber production objectives alone. However, if you want to provide more diverse ground cover for forest-interior birds or ruffed grouse, you may still achieve both timber and wildlife goals. A key benefit of ecosystem restoration as a management objective is in the conservation of rare or critical components of important ecosystems. You need a large land base to accommodate all the components and to create a viable community that achieves these outcomes.

Other objectives

A National Woodland Owner Survey published detailed information about who owns most of the trees in the United States. Sixty percent of the respondents owned less than 10 acres of forest. Most said their land was part of a home or cabin lot, a farm or ranch. Other objectives of ownership included, in order, beauty, nature protection, pass land to heirs, recreation (including hunting), investment, and the production of timber or firewood.

You may see several of your objectives in that list. The ranking of these objectives changed since the previous survey and will probably change again in the future as economic and conservation pressures ebb and flow. Managing for multiple objectives allows you to make subtle shifts in your own personal ranking system because you know what is on your land.

Tax and estate planning are important for many landowners and require careful planning. Because of the depth of those issues and the differences in each state, they are not discussed in this book. Explore these subjects in more detail through other published resources or through experts, such as extension forestry specialists.

Landbase considerations

As mentioned earlier, some land management objectives require larger landbases than others. And when two or more objectives are important, the landbase requirements may increase substantially. In most circumstances, all desired objectives cannot be achieved on 10 acres. Providing adequate early successional cover for a wildlife species in conjunction with mature forest and a wetlands component may encompass more area than you have available.

You need to consider what amenities the lands that surround your property provide. If your main objective is to improve habitat for a certain wildlife species, which component of habitat for that species is most limited on the surrounding landscape? Focusing on providing what is lacking elsewhere will make your efforts more successful.

Land management objectives need to match the conditions of the property. Timber production is not as profitable on rocky ridge tops with shallow soils as it may be on middle to lower slope positions. Sites where fire may not be practical because of risk factors (fire escape or smoke) are not good candidates for ecosystem restoration. However, maintaining oak dominance on these dry sites is relatively easy (Figure 13.3). On the other hand, it may not be financially feasible, or ecologically desirable, to convert coves and stream sides to oak-dominated stands.

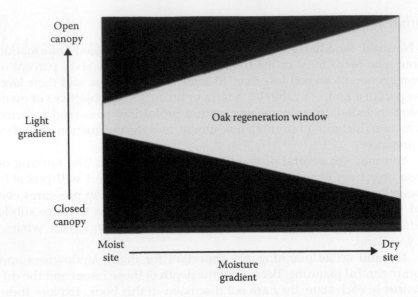

Figure 13.3 The "regeneration window" for oaks in Eastern U.S. forests in relation to light and soil moisture. The window (open area) defines the region most favorable for successful oak regeneration. The region is relatively narrow on moist sites but widens with increasing dryness. On moist sites, intermediate light intensities offer the best opportunities for oak seedling survival and growth. Lower light intensities are insufficient to meet oak's minimum light requirements for photosynthesis but are sufficiently high to inhibit the development of many competitors. As site conditions become drier, the window widens because of the oak's drought tolerance and the exclusion of competitors that are less drought resistant. (Adapted from Johnson, P.S. et al., *The Ecology and Silviculture of Oaks*, CABI, New York, 2010; Hodges, J.D. and Gardiner, E.S., Ecology and physiology of oak regeneration, General Technical Report SE-84, U.S. Department of Agriculture, Forest Service, Southeastern Forest Experiment Station, Asheville, NC, 1993, pp. 54–65.)

Conclusion

As a landowner seeking to manage oak forests in a sustainable manner, you likely will have multiple objectives. You will also have to evaluate the trade-offs associated with the management activities you choose to implement.

Make sure your chosen objectives are compatible with the size and condition of your property. Partner with adjoining landowners to coordinate objectives, timing, and administration of management practices. Work with professional foresters and Certified Wildlife Biologists® to determine the proper management required to meet your objectives over the long term.

Further suggested reading

Aldrich, S.R., Lafon, C.W., Grissino-Mayer, H.D., DeWeese, G.G., and Hoss, J.A. 2010. Three centuries of fire in montane pine/oak stands on a temperate forest landscape. *Applied Vegetation Science* 13(1):36–46.

Birch, T.W. 1996. *Private Forest Landowners of the United States, 1994*, Resource Bulletin NE-134. U.S. Department of Agriculture, Forest Service, Northeastern Forest Experiment Station, Radnor, PA, 183 pp.

Butler, B.J. 2006. Family forest owners of the United States, 2006. General Technical Report NRS-27, U.S. Department of Agriculture, Forest Service, Northern Research Station, Newtown Square, PA, 72 pp.

Delcourt, P.A., Delcourt, H.R., Ison, C.R., Sharp, W.E., and Gremillion, K.J. 1998. Prehistoric human use of fire, the Eastern agricultural complex, and Appalachian oak-chestnut forests: Paleoecology of Cliff Palace Pond, Kentucky. *American Antiquity* 63(2):263–78.

Gingrich, S.F. 1967. Measuring and evaluating stocking and stand density in upland hardwood forests in the central states. *Forest Science* 13(1):38–53.

Guyette, R.P. and Spetich, M.A. 2003. Fire history of oak-pine forests in the Lower Boston Mountains, Arkansas, USA. *Forest Ecology and Management* 180:463–74.

Hodges, J.D. and Gardiner, E.S. 1993. Ecology and physiology of oak regeneration. General Technical Report SE-84, U.S. Department of Agriculture, Forest Service, Southeastern Forest Experiment Station, pp. 54–65.

Hutchinson, T.F., Long, R.P., Ford, R.D., and Sutherland, E.K. 2008. Fire history and the establishment of oaks and maples in second-growth forests. *Canadian Journal of Forest Research* 38:1184–98.

Johnson, P.S., Shifley, S.R., and Rogers, R. 2010. *The Ecology and Silviculture of Oaks*, 2nd edn. CABI, New York.

Luppold, W.G. and Bumgardner, M.S. 2007. Examination of lumber price trends for major hardwood species. *Wood and Fiber Science* 39(3):404–13.

Rodewald, A.D. and Abrams, M.D. 2002. Floristics and avian community structure: implications for regional changes in eastern forest composition. *Forest Science* 48(2):267–72.

Vitz, A.C. and Rodewald, A.D. 2007. Vegetative and fruit resources as determinants of habitat use by mature-forest birds during the post-breeding period. *Auk* 124(4):494–507.

chapter fourteen

How do I manage upland oaks?

David Wm. Smith, James E. Johnson,
John W. Groninger, and Mark E. Banker

Contents

Thoughts about forest management 193
Planning for forest management activities.. 195
Evaluating your forest: The implications of site quality 197
    Topographic and landform features.. 197
    Tree species composition in relation to site quality............................. 199
Your forest: How do you manage it?... 203
Mature mixed-oak forests 203
    Perpetuating the present stand 207
    Regenerate the present stand.. 208
    Continue the present stand with no management 212
Middle-aged mixed-oak forests 214
    Description 214
    Assessing the middle-aged forest..214
    Selecting crop trees.. 215
    Other management considerations and options....................................216
    Summary 217
Young mixed-oak forests.. 217
    Description 218
    Assessing and managing the young forest... 219
    Summary 221
Further suggested reading.. 222

Thoughts about forest management

Following sound management principles can help you maintain and improve upon the many values and uses of your forestland. We want you, the landowner, to have as much information as possible about your unique forest, so that you can make informed decisions that are based on the best science and professional experience available. As you have probably determined by now, forest management decisions have very long-term implications and the exact outcomes of these decisions are, like so many other long-term decisions, filled with a degree of uncertainty. The end result of many forest management decisions can take decades before they are fully realized and can be assessed as to what degree the objectives have been met. Therefore, many forest management decisions reach generations into the future. We also feel that it is important to build flexibility into the plan so that during periodic evaluations appropriate modifications can be implemented. Such things as emerging wood market opportunities, insect or disease infestations, significant ice/snow/wind damage events, new research findings or the need to meet personal financial obligations would automatically trigger an evaluation of the forest management plan. Think of your forest management decisions as having three basic components: biological, economic, and social. The decisions must be biologically possible, economically viable, and acceptable to you and your family. We also think it is worthy to note that if you do not get the biological component right, the economic concerns will not make much difference since the underlying tenets of maintaining stand integrity and ensuring sustainability are unlikely to be achieved.

The preceding two sections and associated chapters have provided a great deal of background information that we hope will be helpful as you focus on your forest management goals and begin formulating management strategies that will satisfy those goals. Again, the more you know and understand about the nature of upland mixed-oak forests, the greater the probability of management success.

In the preceding chapter, a wide range of values and uses for upland mixed-oak forests that you might consider have been discussed in some detail. The specific objectives that you will ultimately work toward will be greatly influenced by the specific physical and biological attributes of your particular forest. In other words, the existing species composition and tree sizes and ages (often referred to as the *stand structure*) and the specific site characteristics have a lot to do with successfully achieving a particular set of objectives (see Chapter 12). It is also important for you to realize that different parts of your forest may support differing objectives. For example, a certain tree species or group of species may have a greater potential for specific birds or animals, whereas other tree species combinations may be highly desirable for high-value forest products or other important uses that are of interest to you. You also should keep in mind that forest structure changes over time. Think about what animal or bird species find excellent food and cover in a very young stand prior to crown closure (see Chapter 4). After 10–15 years, the tree crowns expand and drastically change the light and water conditions and plant composition on the forest floor. As a result, wildlife species associated with the stand change as conditions within the stand change. You need to always keep in mind that a forest is dynamic, *and the only thing constant in a forest is "change."*

Planning for forest management activities

The more you know about your forest and the better you understand how it works, the better you will be able to decide what objectives are compatible with your long-term goals and with the attributes of your particular forest. You need to ask yourself: "What information do I need to know and where might I get that information?" The following list will help you get started. Additional information is provided on this subject in Chapter 18.

1. *Deed to the property*: The deed will contain the land survey calls that identify the property boundaries when plotted on a map.
2. *Topographic maps*: These maps are prepared by the USDI, Geological Survey often at a horizontal scale of 1:24,000 and contour interval of 20 feet. It is most helpful to plot the property boundaries on the topographic map so that important topographic features such as drainages, streams, lakes, ravines, ridges, valleys, coves, roads, pipelines, power lines, and other important features can be identified.
3. *Aerial photographs and other forms of remote imagery*: Most areas have had vertical aerial photographs taken at 10 to 15-year intervals starting in the 1940s. They were often used in developing forest management plans and are very helpful in determining changes in forest cover, harvesting activities, land use changes, and development activities. They are still available at various locations on the Internet and often at local USDA Natural Resource Conservation Service offices. A broad range of satellite imagery is available from a variety of sources, such as Google Earth and other web-based mapping sites, and can be very useful in the management planning process (see Chapter 18 for additional information regarding aerial photography).
4. *Soil surveys*: Soil surveys are maps and associated descriptions that display and interpret soils at any given location. Soil survey information can be very useful in the forest management decision process. This is especially true if your stand has been severely highgraded or mismanaged and the species composition and tree quality no longer represent the true potential of the site, or if the site is presently not forested and you wish to establish a forest. Bear in mind that the soil and associated climate at your location control the forest composition, stand structure, health, and growth at that site. Soil surveys are generally published for a single county. Information about soil surveys is available from your local USDA

Natural Resources Conservation Service office or on their website at: http://websoilsurvey.nrcs.usda.gov/app/WebSoilSurvey.aspx.

5. *Best management practices*: Forestry best management practices (BMPs) are very closely related to soil surveys in that they contain detailed and condition-specific soil and site practices. You should implement BMPs to prevent pollution and protect water quality in streams, ponds, and other waterways that may result from forest management activities. The five types of pollutants are sediment, nutrients, organics, temperature, and chemicals. The primary sources of pollutants on upland sites are sediment and temperature, with organics, nutrients, and chemicals generally being less important. The management activities in your upland mixed-oak forest that most often adversely affect water quality include road building, logging, crossing drainages and streams with logging equipment, and other equipment traffic in and around log decks. It is important that inspections for adherence to appropriate BMPs be made during the harvesting operation so that potential problems are caught and addressed early and damage minimized, as well as upon completion. In addition, BMPs are also applicable in cases where recreational uses such as ATVs, horseback riding, and high-volume hiking create soil erosion problems. BMPs are generally prepared at the state level, and you can obtain detailed information about applicable BMPs from your state forestry agency.

6. *Rare, threatened, and endangered species*: Almost every state has a register or similar publication that provides a listing of and details about rare, threatened, and endangered plant and animal species. These publications will help you identify areas where you might expect to find these important plants and animals. The state forester, state fish and wildlife, or natural heritage agency should be able to provide you with the details on where to obtain information about rare, threatened, and endangered plant and animal species in your locality.

7. *Professional forest management assistance*: As previously stated, the purpose of this book is not to make you a professional forest manager but to provide you with a solid background in the types of information related to forest resources management that are needed for you to ask the appropriate and important questions, and make informed decisions about how to manage your forest property. When you are ready to pursue the development of a forest management plan, we suggest that you seek assistance from a qualified professional forester and other natural resource professionals as may be appropriate to address your proposed management objectives. A list of forestry resources is provided in Chapter 12.

Evaluating your forest: The implications of site quality

Developing an understanding of how well trees and associated plants will grow in your forest is essential to making appropriate decisions about what management objectives are feasible and attainable. We define the term *site quality* as being the innate ability of a site or location to grow things (see the sidebar, "What Is a Site?" in Chapter 3). Just as trees and plants respond differently to site conditions, so do the wildlife species and abundance of those species. In other words, site quality controls what can grow and live on a site. The major factors that control site quality are soil physical and chemical properties, water availability, and the availability of light at various locations within the forest—at the forest floor, in the understory canopy, and in the main canopy. In your garden, it is quite easy to enhance site quality. You do this by watering, adding organic matter, applying fertilizer, controlling insects and disease, and making sure you have plenty of sunlight. In upland mixed-oak forests, these types of activities are, for the most part, not done because they are usually not economically justifiable or physically practical across large forest stands.

In the paragraphs that follow, we will discuss some of the principles and processes that you might consider in evaluating the potential of your forest resources. This evaluation will provide you with baseline information for determining your management objectives and then formulating the forest resource management plan that will attain those objectives and result in the long-term sustainability of your forest.

Topographic and landform features

There are a number of ground-based physical features that you will be able to identify on a topographic map or directly on the ground that will be important in determining your objectives. These features can then be located on the appropriate soil survey map and additional helpful information gained. In general, many of these features may be important habitats important for wildlife and plant species that are not common on the rest of the property. These special places should be identified on maps. The list of special features could include rock outcrops, cliffs, caves, intermittent and perennial streams and drainages, springs, seeps, wetlands, and areas where slopes are in excess of 60%. These special areas are also the locations where you are most likely to encounter rare, threatened, and endangered species. Site-specific management strategies should be formulated to properly conserve these high-value special areas. For all water-related features, you will need to identify appropriate streamside management zones and buffers. Appropriate measures for maintaining the ecological integrity of these areas can be found in BMP guidelines, soil surveys, and similar publications, and from forest resource professionals.

In areas where slopes exceed 5%–10%, the slope percent, the slope position, and the aspect are important landform features. The *slope percent* is the vertical rise in feet per 100 feet of horizontal distance expressed as a percent. An 18% positive slope will rise 18 feet over a horizontal distance of 100 feet. In general, the steeper the slope, the lower the site quality.

You identify *slope position* by referring to the location on the landscape. The *summit* would be the relatively flat upland. As you move to the edge of a slope and start to go downhill, you would be on the *shoulder*. If you move downhill from the shoulder to mid-slope where the slope percent is constant, you would be on the *back-slope*. As you near the bottom of the hill where the slope percent declines and the shape of the slope becomes concave, you would be on the *foot-slope*. As you reach the bottom of the slope and the slope levels off, you would be on the *toe-slope*. From a site quality standpoint, the shoulder would have the lowest site quality, the back-slope would be next, followed by the summit, with the foot-slope/toe-slope having the highest site quality and being most productive (Figure 14.1).

We define *aspect*, often referred to as exposure, as being the cardinal direction that a slope faces. For example, if you are standing on a slope and facing directly down slope and your compass reads N45°E, you would be on a northeast aspect. From a site quality perspective, north, northeast, and east aspects have the highest site quality, followed by southeast and northwest aspects, with south, southwest, and west aspects having the lowest site quality.

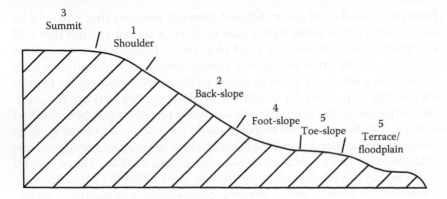

Figure 14.1 Diagram of slope position on the landscape and the relative site quality rank. A rank of 5 is most productive and 1 is least productive.

Tree species composition in relation to site quality

Now that you have some background into what determines site quality, we can proceed with a discussion of what might be the trees species composition in an upland mixed-oak forest on a specific site.

As you have probably gathered by now, the tree species composition of upland mixed-oak forests is highly variable and strongly influenced by site quality. In order for you, as the landowner, to begin to recognize the potential of your forest, we have developed brief generalized forest *Type Group* descriptions along a site-quality gradient for three broad site-quality classes. A *type group* is similar to a *forest type* or *stand type*, terms often used by foresters, but it differs in that it is specifically related to site quality. The *type groups* used in this chapter are specifically defined to describe common oak and associated species that are likely found together along a site-quality gradient on uplands throughout the oak region—less productive, medium or average, and most productive or good sites.

Chestnut oak–post oak type group: The chestnut oak–post oak type group represents tree species you might expect to find on less productive upland sites and includes oak site indices (base$_{50}$) of 60 and less (Figure 14.2). *Site Index* is a species-specific indicator of site quality (or forest productivity) expressed as the average height of the tallest trees in the stand at a specified index or base age. For our discussion of upland mixed-oak forests, the index or base age is 50 years; therefore, a stand with a site index$_{50}$ of 58 is a stand where the average height of the tallest trees is expected to attain 58 feet at age 50. Although these species intermix, chestnut oak is most often found from the Ohio River south and from the Mississippi River east through the Appalachian region onto the Piedmont along the Atlantic Coast states. Post oak is the major oak species on poor, dry sites across the Atlantic and Gulf Coastal Plain through the Ozarks and west to the western edge of the oak region. Bur oak would be common on the less productive sites in forested areas within the Midwestern Prairie region and north of the Ozarks. Commonly associated species would include sassafras, bear oak, chinkapin oak, blackjack oak, blackgum, eastern redcedar, pitch pine, Virginia pine, and shortleaf pine, with red maple and scarlet oak mixed in on the better of these poorer sites. The occurrence of associated species would depend on the native range of the particular species. The presence of the relatively short-lived pine species would suggest that wildfire or other significant disturbance has occurred and allowed these shade-intolerant species to become established. Woody understory shrubs might include blueberry and huckleberry.

Sites where species of this type group are commonly found would be the dryer south and southwest aspects, shoulder and upper back-slope positions, and ridges and uplands with shallow, rocky, sandy, or

Figure 14.2 An example of an upland stand dominated by chestnut oak and scarlet oak (site index$_{50}$ 60). The stand is 80 years old, and the average basal area is 80 square feet per acre.

otherwise nutrient poor soils. You would expect the trees on these less productive sites to be slow growing, relatively short, and often branchy. Regenerating oak species on these sites is generally not a problem following harvest or other major disturbance when there was a significant oak component in the previous stand and where the forest floor is exposed to full sunlight. Stump sprouting is usually prolific with oaks, and competition from non-oak species and herbaceous material is much less than on more productive sites.

High-quality wood products would be limited and are an unrealistic management goal on these sites. However, these drier sites are especially important for various wildlife species watershed protection, aesthetics, and biodiversity, and should biomass fuels or firewood markets be available, the wood would be of significant economic value.

White oak–black oak type group: The white oak–black oak type group represents species found on the medium or average upland sites and includes the oak site indices$_{50}$ of 61–75 (Figure 14.3). Commonly associated species would include the white oak, black oak, scarlet oak, chestnut oak, post oak, pin oak, northern pin oak, northern red oak, southern red oak, bur oak, mockernut hickory, pignut hickory, shagbark hickory, bitternut hickory, blackgum, black locust, American beech, red maple, persimmon, sassafras, eastern white pine, shortleaf pine, and

Figure 14.3 A typical upland stand dominated by black oak, scarlet oak, and white oak (site index$_{50}$ 68). The stand is 75 years old, and the average basal area is 95 square feet per acre.

Virginia pine. A number of these species have natural ranges that do not extend across the entire upland mixed-oak region. For example, you will only find pin oak and northern pin oak in the Central states and across to the Northeastern states, southern red oak from the Ohio River region south, eastern white pine in the Appalachians and to the North, and bur oak from Kentucky and Arkansas north and west. Forests composed of species in the white oak–black oak type group would be the most common of the three type groups.

You would expect to find this type group on the shoulders and upper slopes of north, northeast, and east aspects; on mid-slope, lower-slope, and foot-slope positions of south and southwest aspects; on foot-slope positions of southeast, west, and northwest aspects; and on other gently sloping uplands where soils are moderately fertile and well drained. In the understory, you would expect to find blueberry, deerberry, and huckleberry on the lower-quality sites, and various azaleas, mountain laurel, sourwood, blackgum, greenbrier, American holly, serviceberry, and flowering dogwood on better sites, depending on the geographic location of the stand. The occurrence of pine species (except for the eastern white pine), black locust, sassafras, and persimmon would suggest that fire or other significant disturbance has occurred that allowed these shade-intolerant species to become established.

A wide variety of forest products can be grown on these sites. Under good management, these products would include high-quality sawtimber, pulpwood, specialty products, railroad ties, pallet material, and biomass fuels. Many species of wildlife are found within these stands. A variety of white oaks, red oaks, and hickories helps make these stands important for a variety of wildlife. Biodiversity, aesthetics, and watershed protection are important attributes of these forests.

Obtaining oak regeneration following a harvest or major disturbance is often difficult and requires understanding of silvicultural methods that provide conditions necessary for successful oak regeneration (see Chapter 6). If oak regeneration is not present at the time of the final canopy removal, do not expect oak to be a component of the subsequent stand. It is often necessary to start the oak regeneration process 10 or more years prior to harvest in order to have a reasonable chance of ensuring oaks will be a major component of the future stand.

Northern red oak–mixed hardwood type group: This type group represents species you would expect to find on the best and most productive oak sites in the region and would have oak site indices$_{50}$ of 76 and above (Figure 14.4). You will find combinations of the northern red oak, sugar maple, tulip poplar, black cherry, white ash, American basswood, black walnut, white oak, southern red oak, and American beech on these highly productive sites.

Figure 14.4 An example of a white oak-dominated upland stand (site index$_{50}$ 80). The stand is 85 years old, and the average basal area is 120 square feet per acre. In addition to white oak, the stand contains northern red oak, black oak, hickory, and a few tulip poplar.

These sites would be characterized by deep, well-drained soils that occur on most mid to lower slopes of north, northeast, and east aspects; on lower-slope positions on southeast and northwest aspects; and in coves. On the very best sites in areas of higher rainfall, especially in the Appalachian region, tulip poplar, black cherry, and eastern white pine are extremely competitive and will generally outgrow all of the oak species with the possible exception of northern red oak. The understory will often contain large numbers of the more shade-tolerant species, including the sugar maple, American beech, and red maple.

Because these sites have historically produced high-value products, high-grading and other detrimental harvesting practices have often significantly altered the species composition of these stands to the point that the most desirable and valuable species have been removed, often leaving poor-quality trees and less valuable species. Oak regeneration following any type of main canopy disturbance is extremely difficult to obtain and requires significant site-specific silvicultural manipulations (see Chapter 6).

Your forest: How do you manage it?

Hopefully, the previous sections of this chapter have gotten you in the mood to go out, stroll through your forest, and say to yourself: "I think I am ready to see what I have, use some of the knowledge and tools that I have learned about, and move toward deciding what I want to do." To make this task a bit easier for you, we have developed three categories or scenarios that depict stages of stand development based on tree size and approximate age for you to use in beginning your stand or forest evaluation process. The three categories are *mature mixed-oak forests, middle-aged mixed-oak forests,* and *young mixed-oak forests.* Mature mixed-oak forests are probably the most common and are older forests where the canopy trees average 12 inches or larger in DBH (see the sidebar in Chapter 12 for a definition of DBH) and are greater than 60 years old. The middle-aged mixed-oak forest will have canopy trees that average between 4 and 11 inches in DBH and range from about 25 to 60 years old. The young mixed-oak forest will have trees that average less than 4 inches in DBH and are up to about 25 years of age.

Mature mixed-oak forests

When we think of mature forests, two terms often come to mind: *virgin forests* and *old-growth forests.* It is beneficial to define these two types of mature forests before moving to discussions about the typical second-growth forests that dominate in the upland mixed-oak region. A *virgin forest* has been defined as "an original forest, usually containing large trees, which has not been significantly disturbed or influenced by human activity." These forests also may be referred to as *primary forests*. In the context of our discussion about upland mixed-oak forests, the occurrence of a virgin forest on private lands would be extremely rare and likely nonexistent. Most upland forests were partially or completely cut during the timber boom of the late 1800s and early 1900s (see Chapter 2). Since then, they also have been significantly impacted by several introduced insect and disease epidemics, such as the chestnut blight that wiped out the American chestnut, and the gypsy moth and emerald ash borer that are impacting the structure and species composition of mixed-oak stands today. For these reasons, the virgin forests, as we once thought about them, probably cannot exist today. However, exploring the historical uses of forests in your area can help explain present stand structure, fuel your imagination, and generally enrich your experience as the current owner and manager. An *old-growth* forest on the other hand is a concept and forest condition that could be part of your thinking when deciding on your management objectives. You may not have an old-growth forest or stand right now, but you could have at some point in the future. Old-growth forests or stands are generally characterized by tree age, species composition, canopy structure (layers), presence of dead trees and large branches on the ground, presence of large standing dead trees (snags) in the canopy, large dead branches on the oldest trees, and tree-sized openings scattered in the stand. More specific old-growth attributes for mixed-oak hardwood forests will be discussed in subsequent sections of this chapter.

As a result of the logging history, the majority of today's upland mixed-oak forests are maturing or mature second-growth forests that are more than 60, but less than 160, years old. If your forest fits this general age range, there are three options that you could choose to follow, depending on your long-term objectives. The *first* option is to perpetuate the present stand with specific objectives in mind. These objectives might center on improving wildlife habitat, a long-term goal of developing attributes of an *old-growth* forest or stand, or to improve the quality of the present stand for high-value timber and other wood products. The *second* option would be to start the regeneration process with a goal of growing a new forest. The *third* option would be to let the stand continue on its present path with minimal or no management inputs. Depending on the size of your ownership and the condition of the forest and associated stands, a combination of these options could be appropriate.

With these options in mind, you should evaluate your forest and determine what option or options best fit your objectives (see Chapter 12 for a more detailed description of evaluating forest condition). You should determine the following:

1. *What is the species composition and how are these species distributed in the stand?* The species composition will help you determine what management objectives are viable.

2. *What is the site quality?* If your objectives include high-value wood products, you need to be on the better medium and on the highest quality sites. On the more productive sites, growth rates are good, species composition includes the most desirable species, and the greater stand density fosters branch-free boles needed for high-value sawtimber and veneer. The economic viability of implementing intensive forest management practices on higher-quality sites is usually enhanced because of the higher value of the end products and the shortened time it takes for the trees to reach merchantable size.

3. *What is the overall health of the stand?* Are there a lot of dead and dying trees, or trees with a lot of dead branches in outer portions of the crown? Are there certain tree species where these problems are more prevalent? Are there nonnative invasive trees, shrubs, or vines present in the stand or in adjacent stands—trees such as tree-of-heaven, mimosa, paulownia, chinaberry, or Russian olive, shrubs such as autumn olive, multiflora rose, or privet species, and vines such as various honeysuckle species, oriental bittersweet, or kudzu? In general, you should determine the strategies to control or eradicate these invasive species as soon as possible and prior to canopy disturbing activities.

4. *Are there signs of deer browsing or other predation problems?* You will want to know if the deer population is dense enough to have a significantly negative impact on new regeneration, either in advance of a harvest or after a harvest. Chapter 17 discusses this problem in detail and provides several suggestions for addressing impacts from white-tailed deer.

5. *Are there sub-canopy layers of small trees in the stand and what is the species composition of these layers?* In order to successfully regenerate oaks, the oak regeneration must be free to grow. Most of the oak species are intermediate in shade tolerance (see Chapter 6). Oak seedlings need near full sunlight in order to survive and grow competitively with associated species. In maturing and mature stands, one or more sub-canopies often form under the main canopy.

Figure 14.5 White oak seedlings in the first growing season following a bumper acorn crop. In this growing season, there were 30,000–40,000 seedlings per acre. The stand was about 80 years old with a basal area of 110 square feet per acre and a site index$_{50}$ of 70. Three years later, the stand basal area was about the same; however, the number of oak seedlings had dropped to less than 100 per acre and were in a noncompetitive position on the forest floor.

In most cases, the midstory canopies are composed of shade-tolerant species such as the sugar maple, American beech, American holly, serviceberry, flowering dogwood, or red maple. Any new oak seedlings that establish following a good acorn-producing season have little chance of surviving longer than the initial growing season under the intense shade of these more tolerant species (Figure 14.5).

6. *What is the stand density or stocking level?* When we talk about *stand density* or *stocking level,* we are referring to a measure of crowding. How near or how far a tree is from its neighbors has a significant effect on its shape, health, and how fast it will grow. If trees in a mature stand are very close together, they will tend to have tall straight stems, few branches below the crown, smaller diameters, and smaller crowns than trees under similar conditions where trees are farther apart and the stand is less dense. In the less-dense stand, trees would tend to have greater diameters, perhaps have a few more branches on the bole, and the crowns would be larger. In a stand that is very dense, there is little space and resources available for additional growth,

and as a result trees in the least competitive position (often sunlight is the limiting factor) will die as a result of severe competition. For additional information on stocking, see Chapters 12 and 13.

Perpetuating the present stand

This option would be viable if the present stand is 60–100 years of age, it is your desire to keep the stand for perhaps 40 or more years, the species are considered medium to long lived, and contains species that will meet your objectives. Should your stand contain a significant proportion of relatively short-lived oak species, such as scarlet oak, southern red oak, post oak, or black oak, these species are not likely to survive or remain viable canopy components. If the stand is fully stocked or overstocked, has a good mixture of short- and long-lived oaks, and high-value trees are present, you can consider management strategies that include producing high-value timber products, maintaining abundant hard mast for wildlife, and/or developing an old-growth stand structure.

To develop high-value sawtimber, it is important to maintain a consistent diameter growth rate and ensure branching does not occur on the first 35–40 feet of the bole. If the density increases into the overstocked range (above 120 square feet per acre), the diameter growth will slow. If the density approaches the understocked range (below 70 square feet per acre), branching on the lower part of the bole is likely to occur. Maintaining the stand at the desired density to achieve high and consistent growth and ideal bole form can be accomplished by thinning, improvement cuts, or a combination of the two.

If producing consistent crops of hard mast is an objective, it will be important to maintain mast-producing trees in a dominant crown canopy position, develop large crowns, and keep as much diversity of mast-producing species as possible. Maintaining reasonable spacing between trees where the crowns of desired mast-producing trees are free to expand is important. Since the frequency and volume of acorn production vary among oak species (and even within species) and are not predictable, it is important to maintain a diversity of oak species (both white oaks and red oaks) within the stand, if possible. Keep in mind that acorns of the white oaks germinate in the fall, whereas those from the red oaks germinate in the spring, and that white oak acorns mature in 1 year, whereas the red oaks require 2 years. Other hard mast-producing species, including, hickory, American beech, and black walnut, can be important for wildlife in years of poor acorn production.

Developing an old-growth stand requires a different strategy. An old-growth structure includes such attributes as standing dead trees, large down woody debris, and openings sufficiently large to promote regeneration of a variety of tree species that are randomly distributed throughout the stand—openings that are perhaps 1/10th to 1/5th acre in size would be in the range common for eastern hardwood forests. The structure of old-growth stands is highly variable across the region and is dependent on disturbance timing and intensity. Some old-growth stands will develop what is known as an *uneven-aged structure*, meaning the stand contains trees spanning a range of age classes. The younger trees occur within the openings mentioned earlier, with the older mature trees comprising the surrounding intact forest. This *patchwork* of small openings cycles through time as disturbances continue to create new patches, while trees in the existing patches grow back into the canopy. Natural disturbances that create these conditions obviously take time. If the old-growth condition that you desire is to be characteristic for upland mixed-oak stands, then it is essential that you have the long-lived oak species, including the white oak, chestnut oak, bur oak, or northern red oak, present throughout the stand. In the absence of these species, you would expect the stand to shift toward a species composition where a combination of the sugar maple, hickories, and American beech would eventually prevail. Allowing stand density to increase to the point where trees die because of overcrowding would initially result in the dead trees being standing snags, but in time they would fall and become large down woody debris. As the trees get bigger, some will die and eventually fall, perhaps taking a nearby tree with it, thereby producing the opening that helps create the old-growth condition. In maturing or mature upland mixed-oak stands, we would look at the process for developing old-growth conditions to take perhaps 100 years or more.

Regenerate the present stand

There are a number of reasons why you may want to regenerate the present stand. These reasons would include the following: the present stand condition is such that additional time will not enhance the value of the stand; a major disturbance such as an ice storm, tornado, hurricane-level winds, or wildfire has all but destroyed the stand; or an insect or disease attack appears to be taking place or is imminent. It is also possible that you have a need to generate income; however, the short- and long-term economic implications of a harvest to meet the income need should be carefully evaluated. When you have made the decision to regenerate the stand, one of the first questions you will want to ask is: "Where is the oak regeneration for the next stand going to come from?" If an inventory of the forest floor fails to show that you have sufficient advance regeneration present to establish the new stand, then a main harvest should not be planned until the required regeneration is in place. Oak regeneration challenges are discussed in detail in Chapter 6. Ideally, the oak regeneration establishment process starts a number of years prior to the time that the main harvest occurs. Silvicultural operations such as a thinning or an improvement cut, which reduces stand density, often stimulate oak regeneration by providing the necessary light at the forest floor that will allow oak seedlings to develop into advance regeneration. In general, only advance oak regeneration (stems that are 4 or more feet tall) and not competing with other sub-canopy species will be in a position to be competitive in the new stand when the overstory is removed in the final harvest cut (Figure 14.6). These operations may take place 10–20 years before the final harvest cut. Disturbances such as wind and ice damage, insect- or disease-caused mortality, or fire also may open the canopy and stimulate oak reproduction. It is important to keep in mind that successful oak regeneration almost always comes from advance regeneration or from stump sprouts. Oak regeneration from stump sprouts most often occurs on the less productive sites where the *chestnut oak–post oak* type

Figure 14.6 This 4-foot-tall northern red oak advance regeneration seedling sprout, on a medium site (site index$_{50}$ 70), is in a competitive position should the overstory canopy be opened or removed.

group is prevalent. On these sites, competition from other tree species is not great, and the trees are slower growing and not usually large in diameter. Oak stump sprouting often does not occur on older, larger-diameter, and faster-growing trees associated with the more productive sites typical with the *white oak–black oak* and the *northern red oak–mixed hardwood* type groups. However, when stump sprouts do occur on these sites, they are very important in maintaining the oak component in the new stand. Remember that stump sprouts, regardless of site quality, are generally fast growing and very competitive with regeneration from other tree species.

Before we discuss regeneration alternatives, we need to reiterate the need to identify potential problems with the nonnative invasive and undesirable native species previously mentioned. Any time there is a disturbance in the stand or if the stand is next to an open field or area that has been recently disturbed, these species have an opportunity to invade and become established. Since many of these problem plants thrive under high light availability, we recommend controlling them prior to any major canopy or forest floor disturbance. Details on how to treat these undesirable plants are discussed in Chapter 10.

Several regeneration methods might be considered depending on the condition of the present stand. Of primary importance are the presence, abundance, and distribution of advance oak regeneration, as previously discussed. When a major disturbance occurs, whether it is natural or human caused, there will be a rapid reoccupation of the site by woody vegetation during the following growing season. What is less certain is whether the regeneration will contain the desired oak composition. For the most part, an even-aged or perhaps a two-aged regeneration method likely will be most successful. These would include variations or combinations of the clearcut, shelterwood, and irregular shelterwood (also referred to as leave tree or deferment cutting and result in a two-aged stand structure) regeneration methods. Group selection, an uneven-age regeneration method, has been considered for application in upland mixed-oak forests; however, there are few situations where this regeneration method will result in or achieve the desired oak species composition. Chapter 7 discusses these regeneration methods and the techniques for using them.

If your regeneration survey indicates there is sufficient free-to-grow advance oak regeneration in the understory, then the clearcutting regeneration method would be appropriate. However, should there be concerns about the visual integrity of the stand following the harvest, a desire to sustain some hard mast production, or a goal of producing some large-diameter trees, then you may wish to consider the irregular shelterwood regeneration method that would result in a two-storied and two-aged stand. If the clearcut method were implemented, all trees down to 1 or 2 inches in diameter would be cut with the exception of dead snags or a couple of larger den, cavity, or nesting trees per acre should they be desired for wildlife purposes. The goal is to open the forest floor to full or nearly full sunlight. General guidelines for the irregular shelterwood would be to leave no more than 20 long-lived, well-distributed, main canopy trees per acre of the desired species. These residual trees should be greater than 12 inches in diameter, and their basal area should be between 15 and 20 square feet per acre. There is no intent for these leave trees to serve as a source of regeneration for the new stand; however, should species such as tulip poplar or sugar maple be left as residuals, you would expect regeneration of both species to occur from new seeds and tulip poplar regeneration from seeds stored in the forest floor.

In the event your regeneration survey indicates there is insufficient advance oak regeneration to achieve your oak component objectives, some form of shelterwood regeneration method is probably the only viable alternative. The objective of the shelterwood is, through one or more partial tree harvests or removals, to provide the forest floor conditions that will result in the establishment, growth, and development of the desired oak advance regeneration. In most cases, this process would include the removal or killing of non-oak species in any of the sub-canopies. Ideally, this initial process would be implemented immediately following a good or bumper crop of acorns from most of oak species present in the stand (Figure 14.7 or 14.8). In addition, should stand location and other conditions allow, the use of prescribed fire to control understory vegetation is a tool that may be considered (see Chapter 9 for additional information on the use of prescribed fire). Should the removal of sub-canopies fail to provide sufficient light for advance oak reproduction development, a partial removal of the main canopy may be required to reduce the stand basal area to the desired level. In most cases, a basal area of around 60 square feet per acre should be sufficient to foster oak reproduction growth and still limit the growth of some of the faster-growing, shade-intolerant species common in these stands. The success of the shelterwood method depends on the timing and amount of the acorn crop with regard to the canopy opening process. Failure to obtain the needed advance regeneration following the initial treatments likely will require additional sub-canopy and overstory manipulations. The time required from the initial treatment to the point that sufficient advance regeneration is achieved and the final harvest planned is often 5–15 years.

Even-age and two-age regeneration methods that use natural regeneration to establish the new stand are by far the most common and effective methods for regenerating eastern upland mixed-oak stands. However, artificial regeneration of oaks should be mentioned as having some applicability under special circumstances. For additional information on artificial oak regeneration, refer to Chapter 8, and for a discussion on developing an uneven-age structure, see Chapter 7.

Figure 14.7 Scarlet oak advance regeneration 3 years following a shelterwood cut that reduced the stand basal area to 70 square feet per acre. The stand is 75 years old with a site index$_{50}$ of 70.

Once the final regeneration harvest is completed, your stand will have transcended to the *young mixed-oak* stand described later in this chapter. As a final step, it is important to conduct a final inspection of the harvested area to ensure terms of the contract have been satisfied and the site is free of equipment and other human-made debris. You should make sure all trees designated for harvest have been cut and that appropriate BMPs have been implemented.

Continue the present stand with no management

We present the option of perpetuating the present stand with no management inputs only as a temporary decision or option by the landowner.

Figure 14.8 Scarlet oak advance regeneration 3 years following a shelterwood cut that reduced the stand basal area to 65 square feet per acre. The stand is 80 years old with a site index$_{50}$ of 70.

Our view is that your forest is a "real" asset and an integral part of your financial portfolio. We would agree that you may not have to do anything right now to manage your forest asset. One nice thing about most forests is that there is generally a "window of opportunity" when it comes to implementing forest management. This window is not a matter of hours or days as is often the case in the financial market, but usually is many months and perhaps even a couple of years. Not counting major catastrophic events such as tornadoes or wildfires, changes in the forest are relatively slow. If a market is not good right now, it is often possible to wait months or even a few years for that market to improve or a new market to emerge—the forest will still be there. The exception to this is if exotic invasive species are a significant problem, delaying corrective action could make the eventual cost of treatment considerably higher.

With these conditions in mind, we believe the option to "continue the present stand with no management" is valid only if that decision is governed by two constraints: (1) The decision for no management is made only after a complete stand/forest evaluation has been completed and discussed with a resource professional. In other words, you should make an informed decision. (2) The decision for no management is reevaluated with revised forest inventory information every 5 years or sooner if a major disturbance has occurred.

Middle-aged mixed-oak forests

Managing middle-aged stands can be a very satisfying task for landowners. Stands in their middle years are characterized by trees that are not yet at their full potential to produce mast, are too small to be highly valuable as timber, and are not yet massive and visually striking. In the middle years, you do not need to worry about regenerating the next stand yet and it is too late to plant anything in the understory and expect it to survive. What you do have is the potential to steer the stand into a condition where it can better meet your management objectives down the road. As always, caution needs to be used when embarking on forest management to make sure your actions do not have negative consequences during this transition phase of forest growth. In this section, we present some guidelines for managing middle-aged forests.

Description

A "middle-aged" oak stand is approximately 25–60 years old and characterized by "pole-sized" trees 4–11 inches in diameter. The canopy is almost always closed with very little sunlight reaching the forest floor and very little growing in the understory as a result.

Most middle-aged stands vary in density, requiring attention in some areas and not in others. For uniformly dense stands or dense clumps within stands, the primary management option is removing some trees to make more room for others to grow (see Chapter 11). This is commonly called "thinning," "crop-tree release," or "crown touching release." Here, we will refer to this practice as crop-tree release, with "crop" referring to future timber harvest.

Although middle-aged oak forests support some generalist songbirds and other wildlife, they are much less useful to wildlife compared to young or mature forests. Normally, middle-aged oak provides little food or cover. Nonetheless, some of the more precocious oaks may have begun to produce acorns that may be consumed by wildlife. Carefully removing competitors from seed-producing trees can help the crowns of these trees expand and generate more seeds. It also admits more sunlight to the forest floor, increasing production of woody and herbaceous wildlife foods and enhancing cover.

An important aspect of managing middle-aged oak forests is aiding the progression to well-formed mature stands. This will happen naturally without any management, but the stand will likely reach maturity more gradually.

Assessing the middle-aged forest

Perhaps the most crucial aspect of managing middle-aged oak forests is identifying the trees within them that have the most potential to produce valuable timber and produce the most acorns in the next several years and beyond. Judicious management of oaks at this stage can have a long-term impact on wildlife and timber production. Therefore, at this stage of stand development, we focus most of our attention on individual trees, including how to identify the individuals you want to keep and those that can be cut to improve the stand. If done properly, middle-age stand management will benefit long-term timber value, enhance wildlife habitat, and make the stand healthier and more diverse.

Understanding the growth potential of your forest stands will help you meet different objectives. On very productive sites, oaks are often overgrown by faster-growing species like tulip poplar and red maple. Here, you may want to focus your management on the most potentially valuable crop trees. Less desirable or subdominant competing trees can be removed from around crop trees if they are suppressing crown expansion and, therefore, slowing potential growth. In these stands, oaks may represent only a small portion of the stand. Management can be focused on retaining the best oak trees. On highly productive sites, wildlife-friendly vines such as grape that can affect timber quality may be less tolerable.

In stands of intermediate productivity, consider balancing wildlife and timber objectives by leaving some clumps of vines in low-quality trees while releasing the best trees as crop trees. When in doubt, focusing only on releasing crop trees is usually both an economically and ecologically sound strategy.

On very dry or rocky sites, which often favor oak, trees sometimes do not grow tall and/or are poorly formed (not straight and tall). Timber production may not be a realistic goal on such sites. On these less productive sites, crop-tree release may proceed with maximizing wildlife benefits as a primary objective. If there are few obviously superior oaks, then thinning non-mast-producing species, such as birch, maple, and sweetgum, in favor of mast producers, such as oak, cherry, and hickory, can be a wildlife-friendly strategy. Here, you should encourage vines, such as grape and greenbrier, that add considerably to wildlife habitat quality.

Selecting crop trees

The most important criterion for releasing a crop tree is whether the physical effort and sacrifice of other trees will provide benefit. Thus, you want to make sure the crop trees you select are superior. Most people equate tree size with age. Most trees in a stand will be the same age but may vary markedly in size. The difference in size among trees of the same species may be genetic. We want to select the best genetics for our crop trees.

A common mistake is releasing crop trees when they are too young. When you remove competitors from around a healthy tree, its branches tend to become larger and natural self-pruning (lower branches die and fall off) may cease. These branches may interfere with the development of straight, branchless tree trunks that later can mean a more valuable tree if you choose to sell the trees. A common rule of thumb is to wait until a tree is about 35–40 feet tall before releasing it. Although we are focusing on oaks, avoid the temptation to release an oak that is not healthy looking if timber production is your objective. For additional information on thinning and releasing crop trees, see Chapter 11.

Dead trees (snags) are common in a middle-aged stand. They do not use any resources and do not compete with living trees, so there is no need to remove them. Snags in stands this age have limited wildlife value because of their small diameter. Nonetheless, they may provide cover for some small mammals and herpetofauna as woody debris on the forest floor.

Aesthetics may be a consideration. Interesting or attractive trees often do not have the characteristics of a potentially valuable timber tree. Some aesthetic traits may include the following:

- Species with attractive blooms or fall foliage (dogwoods, maples, serviceberry, etc.)
- Individuals in high visibility areas like along streams or roadsides or from old pastures and fence line trees
- Other interesting features such as unusual bark or multiple trunks

Sometimes attractive trees are bent, forked, or branchy, but do not be afraid to select these as crop trees to give your forest stand or property some added character.

Other management considerations and options

Vine control: Grapevines can be a problem in some of our most productive hardwood forests. Tree crowns containing vines are weakened and more susceptible to snow and ice damage. Vines compete directly with the host trees for light, water, and nutrients. Vines should be addressed before crop-tree release is practiced. Cutting vines at the base is usually sufficient for control. For best results, manage vines a few years before thinning to allow for follow-up treatment prior to adding sunlight. If wildlife management is an objective, then you may choose to leave the grapevines, as well as various other species, because of the food and cover value they provide. When meshing wildlife objectives with timber production objectives, you may leave grapevines in patches, preferably in small, subdominant trees. Some of these trees can be cut to create grape tangles on the forest floor and in the shrub layer that provides cover for wildlife at ground level.

Invasive plant control: Increasingly, invasive nonnative plants are gaining a foothold in our oak forests. Japanese, Tatarian, and Amur honeysuckle, privet, Japanese stiltgrass, oriental bittersweet, autumn olive, barberry, and garlic mustard are just a few examples. Because these species typically benefit from any forest operation that adds sunlight to the forest floor, it is critical to remove invasive nonnative species prior to any management activity. Watch for new nonnative plants once you have added sunlight and remove them immediately either mechanically, with herbicide, or both.

Fire: Low-intensity prescribed fire may be used in middle-aged stands to enhance the structure and composition of the understory for wildlife. However, some type of thinning is required to provide sufficient sunlight and stimulate the understory. If your objective is timber production, fire is not recommended at this stage of stand development.

Insect and disease control: Specific threats to oak forests include oak wilt, gypsy moth, and oak decline. Susceptibility to each of these varies, and the presence of one or some of these as well as other threats may impact your stand this year, next year, or sometimes several decades from now. In the face of this uncertainty, managing for a vibrant and healthy forest is your best defense. When thinning trees, be sure to maintain a good mix of species—it is unlikely that one threat will affect all species equally.

Summary

Middle-aged stands represent the period when the strongest, genetically superior trees start to become apparent, and you can help facilitate their development and ensure the maintenance of your oak component. Focus your energy and resources where they will do the most good. Provide growing space for the trees most likely to provide wildlife and commercial benefits via release.

Fortunately, there are many dedicated and talented foresters who can help you avoid mistakes during this transition stage. You might also seek out a landowner who has done similar work and visit their woods to see what you might want to work toward. Once you get comfortable with the procedures and potential outcomes, remember to think long term, be safe, and have fun!

Young mixed-oak forests

One of the exciting things about a young mixed-oak stand is that when compared to the "mature" forest, things seem to happen quickly—you can actually see things grow and happen over a period of months. As soon as the overstory is removed, whether it is by a final harvest or from a major natural disturbance, the forest floor comes alive and the process of ecological succession is on its way to establishing your next forest.

Description

A "young" oak stand is characterized as being less than 25 years old and containing tree seedlings, seedling sprouts, and saplings less than 4 inches in diameter, and a variety of other vegetation depending on the age and growth rates of the woody material. Throughout much of the upland oak region, this age forest is the least common, yet extremely important for maintaining wildlife diversity. Many wildlife species require young forests (Figure 14.9) for nesting and foraging. Soft mast, primarily blackberry, blueberry, and pokeweed, that occur the first few years after a harvest is important for a wide variety of wildlife species. In addition, insects and

Figure 14.9 A young, well-stocked, mixed-oak stand dominated by oak stump sprouts and seedling sprouts 6 years after a clearcut regeneration harvest on a medium to poor site (site index$_{50}$ of 65).

fungi from rapidly decaying woody and other organic material provide valuable food and cover for many wildlife species.

If your stand is 1 to 5 years old, you should expect to see tops and branches from the harvest, a thicket of underbrush, including various herbaceous species, blackberry, raspberry, blueberry, huckleberry, deerberry, perhaps sumac, redbud, serviceberry, and a large number of future main canopy tree species. When you first see a young stand in this condition and if your thoughts are something like: "I don't think that I want to walk very far into this area right now," then your new stand is looking like it should. You also may think it is not very aesthetically pleasing, but ecologically it is wonderful. At this young age, species such as the field sparrow, common yellowthroat, golden-winged warbler, American woodcock, Appalachian cottontail, snowshoe hare, wild turkey, and white-tailed deer will be using the low vegetation for nesting, cover, and forage.

If your stand is between 5 and 15 years old, the herbaceous groundcover is declining rapidly, if any is still left, as the tree saplings develop crowns that exclude light from the forest floor. The shade-intolerant shrubs also will be less vigorous and declining. During this span, the young forest will become prime habitat for ruffed grouse, chestnut-sided warbler, and yellow-breasted chat and will continue to be used extensively by deer, elk, cottontails, hares, and even bears.

As your forest progresses toward the *middle-aged mixed-oak forest*, the crown canopy has closed and the herbaceous cover on the forest floor largely will have disappeared except for the most shade-tolerant species, such as a variety of ferns.

Assessing and managing the young forest

There are three likely pathways by which you obtained your young mixed-oak forest. The first might be that you have owned the property for some time and have just completed a major planned harvest that opened the forest floor to full sunlight. If this is the case, we would expect that in preparing for the final harvest, you took the management steps necessary to ensure that there was sufficient oak advance regeneration in place to meet your objectives. Under this scenario, there is little you probably will have to do until the main canopy grows, closes, and the stand moves into the middle-aged category, other than to be watchful for possible excessive deer browse and the presence of nonnative invasive species. Chapter 17 discusses excessive browsing by deer, and you should seek advice from a wildlife specialist if it becomes a problem. If there are unwanted tree species, shrubs, or vines present that are likely to materially affect the growth of your desired oak species, control actions are probably warranted. This would be especially true if the undesirable trees are in a dominant position or the vines are in the crowns of the desired species. Control would likely be with a foliar or stem herbicide treatment or a combination of cutting and stump herbicide treatment (see Chapter 10). A common example of this would be red maple sprout clumps that originate from cut stumps. These clumps are fast growing, usually have multiple sprouts (10 or more) that tend not to self-thin, but rather increase in size, take up a large amount of growing space, and outcompete most oak regeneration. The sooner you can accomplish this control, the easier it will be. On better sites, crop-tree release (see Chapter 11) could be an option for improving growth, stand quality, and even wildlife habitat.

The second possible pathway for having a *young stand* is that you own the stand and a major natural disturbance, such as a large gypsy moth outbreak, has partially or completely destroyed the main canopy of the stand, or the event will lead to the loss of the main canopy in the very near future. The probability of this happening in your stand is remote, but it is worth a brief discussion. Severe ice or snow storms, hurricane or tornado force winds, wildfire, and insect or disease epidemics will be the most likely causes of such a disturbance. In the event that your oak or mixed-oak hardwood stand has been severely impacted by gypsy moth related mortality, keep in mind that tulip poplar is very resistant to the gypsy moth and red maple to a lesser extent. This will result in both these species being much more dominant in the future stand. We highly recommend that you seek the assistance of a professional forester to properly evaluate and gain advice on appropriate actions to minimize future unintended consequences of the event. It is likely that some sort of salvage operation would be in order. In most cases, trees with broken tops, those that are partially uprooted, or are heavily damaged should be cut and salvaged if possible. If they cannot be salvaged, they should still be cut but left on the ground. By cutting them you will take advantage of sprouting as a source of regeneration. For the most part, the new stand will be composed of the advance regeneration that was in place prior to the event, and seedlings developed from seeds stored on the forest floor, seed from the tops of downed trees, and stump sprouts from cut trees. Potentially viable oak regeneration would probably come from the advance regeneration pool and from stump sprouting. New oak seedlings that come from the present-year acorn crop are very unlikely to be competitive, except perhaps on the least productive sites. You may wish to consider the use of artificial regeneration, either by planting seedlings or by direct seeding. It is important, however, that prior to committing to this endeavor, you fully evaluate the cost and the probability for success involved with artificially regenerating oaks.

The third pathway by which you may have obtained your *young stand* is that you purchased the land after a final harvest or a major natural disturbance. As in the second scenario, you should seek advice and counsel of a professional forester and conduct an appropriate stand evaluation. The important things to determine include the amount and distribution of oak advance regeneration and oak stump sprouts, the presence, abundance, and distribution of invasive or undesirable trees, shrubs, and vines, and whether browse damage from deer is a significant problem. Armed with the stand evaluation, you are in a position to take any appropriate management action discussed in the first two scenarios. It also is possible that your recently purchased young stand is the result of a "high-grading" or "diameter-limit" final harvest and there is a significant amount of poor-quality trees, undesirable species, or severely damaged stems remaining on the site. It is not uncommon to have basal areas of 10–40 square feet per acre of undesirable material present following this type of harvest. Under these circumstances, it is very important to reduce the basal area of the undesirable stems to less than 5 square feet per acre. This would include oak species which, if cut, would likely develop stump sprouts that would be potential desirable crop trees in the future. It is also important to keep in mind that tulip poplar and red maple often become dominant species in these stands because of tulip poplar's ability to develop highly competitive seedlings from seeds stored on the forest floor and from tree crowns, especially on higher-quality sites, and red maple's tendency to develop sprout clumps from cut stumps.

Summary

When compared to the active management activities and the relatively large trees we described in the *mature* and *middle-aged* forest discussions, the amount of necessary management activity in your *young forest* may be relatively small. The one exception to this statement is that because the stand is young and the trees are small, it is relatively easy for you to use your spare time to control species composition without the need for special skills (other than being able to identify the species), or needing expensive, specialized, or heavy-duty equipment. Just a few hours each week can make a big difference in the future, and it is fun to know that you helped shape tomorrows forest.

There is a lot going on in the *young forest*; everything is growing, new plant species appear and then disappear, animal and bird populations flourish, and change is rapid. Initially, you may want to just sit back, watch, and enjoy your forest. At first you will probably walk around it. Then, as your interest gets the upper hand, you may want to have a part in the change that is occurring, and in a couple of years, you will enjoy walking through it knowing that you had a part in growing your forest.

222 *Managing Oak Forests in the Eastern United States*

Further suggested reading

Barrett, J.W. (ed.). 1994. *Regional Silviculture of the United States*, 3rd edn. John Wiley & Sons, Inc. New York, 643 pp.

Brose, P.H., K.W. Gottschalk, S.B. Horsley, P.D. Knopp, J.N. Kochenderfer, B.J. McGuinness, G.W. Miller, T.E. Ristau, S.H. Stoleson, and S.L. Stout. 2008. Prescribing regeneration treatments for mixed oak forests in the Mid-Atlantic Region. General Technical Report NRS-33, USDA Forest Service, Newton Square, PA, 100 pp.

Burns, R.M. and B.H. Honkala (eds.). 1990. *Silvics of North America, Vol. 2: Hardwoods, Agriculture Handbook 654*. USDA Forest Service, Washington, DC, 877 pp.

Helms, J.A. (ed.). 1998. *The Dictionary of Forestry*. Society of American Foresters, Bethesda, MD, 210 pp.

Nyland, R.D. 2001. *Silviculture Concepts and Applications*, 2nd edn. McGraw-Hill Companies, Inc., New York, 682 pp.

Roach, B.A. and S.F. Gingrich. 1968. *Even-Aged Silviculture for Upland Central Hardwoods, Agriculture Handbook 355*. USDA Forest Service, Washington, DC, 39 pp.

Smith, D.M., B.C. Larson, M.J. Kelty, and P.M.S. Ashton. 1996. *The Practice of Silviculture: Applied Forest Ecology*, 9th edn. John Wiley & Sons, Inc., New York, 537 pp.

Spetich, M.A. (ed.). 2004. *Upland Oak Ecology Symposium: History, Current Conditions, and Sustainability*, October 7–10, 2002, Fayetteville, AR. General Technical Report SRS-73, USDA Forest Service, Ashville, NC, 311 pp.

chapter fifteen

How do I manage for woodlands and savannahs?

Patrick D. Keyser, Craig A. Harper, McRee Anderson, and Andrew Vander Yacht

Contents

What are woodlands and savannahs?...224
What has become of woodlands and savannahs?....................................226
What are some of the benefits of woodlands and savannahs?...............229
Restoring woodlands and savannahs 232
    Step 1: Pick a site 232
    Step 2: Get rid of unwanted trees..233
    Step 3: Burn 238
A word about fuel loads...240
A word about herbicides..240
A word about invasive plants 240
When am I done? 241
Regeneration..241
Let's get started! 242
Further suggested reading...245

Prior to European settlement of Eastern North America, fire was a common phenomenon in what we now think of as our oak forests. Native Americans routinely used fire to clear land for raising crops, improve range for game, facilitate hunting, and protect themselves from other tribes during times of conflict (see Chapters 2 and 9). Those frequent fires, along with numerous fires ignited by lightning strikes, maintained prairies and savannahs for millennia, restricting thicker oak forests to damper sites where fire was less frequent. As ridges and rivers encouraged or discouraged the spread of fire, the mix of prairie and forest occurred in greater or lesser portions, respectively. In general, moving from west to east, the patchwork quilt of land became progressively more forest dominated.

Early settlers left numerous accounts that describe open, park-like forests, and even prairies, in what are now Virginia, North Carolina,

Tennessee, and Kentucky. Andrew White, a Jesuit priest on an expedition along the Potomac in 1633, observed that the forest was "not choked with an undergrowth of brambles and bushes, but as if laid out by hand in a manner so open, that you might freely drive a four horse chariot in the midst of the trees." Bison and elk were once abundant in these same areas, a testament that grass-dominated groundcover was common in the region. One early traveler's account of a scene on the now heavily forested Cumberland Plateau in Tennessee observed in 1783 that, "The top of the mountain... being... a vast upland prairie covered with the most luxuriant growth of native grasses, pastured over as far as the eye could see, with numerous herds of deer, elk, and buffalo...." Relentless and often wasteful harvesting of these native herbivores from the 1600s into the early 1800s resulted in the complete decimation of the region's herds. And as settlers moved westward, burning patterns associated with the native people began to fade, as did the grasslands that depended on fire.

What are woodlands and savannahs?

The transition between prairie and forest is gradual and a result of the mixture of forces that create both the prairie and the forest: precipitation, fire, and grazing. To the West, there is less moisture and more fire, to the East, more moisture and less fire. Historically, grazers such as bison and elk followed the grass and slowed the change back to more forest-like conditions through the disturbances they caused.

Biologists describe an area where there is a widely spaced, open canopy, with grass and forb groundcover, as a savannah (Figure 15.1a). A woodland is a gradation between savannah and forest: a heavier tree canopy, sparser herbaceous, and more woody cover in the ground layer, but still with a herbaceous understory (Figure 15.1b). Woodlands become forests when canopy closure is virtually complete and woody rather than herbaceous species dominate the understory. Savannahs typically are described as having 10–35 square feet of basal area/acre of overstory trees (10%–30% canopy cover), whereas woodlands typically have 40–65 square feet (40%–60% canopy cover, in some cases, up to 80%). Typically, savannahs would have been exposed to more burning than woodlands, resulting in more open canopies and well-developed herbaceous ground layers.

The species of overstory trees really does not matter, though they generally are fire tolerant by necessity. Although few woodlands and savannahs exist today, there are enough remnants to get a good picture of how they would have looked historically. From the southern edge of the Lakes States southward to the Ohio Valley, the dominant species were oaks. From the Ohio Valley to the Fall Line (the beginning of the Coastal Plain),

(a)

(b)

Figure 15.1 Typical oak savannah (a) and woodland (b) with open overstories allowing for the development of an herbaceous ground layer. Note the more open canopy in the savannah. (Photo courtesy of C. Coffey.)

oaks were mixed with shortleaf pine, with soils, slope, and aspect, as well as site history determining which dominated. Further south, oaks and shortleaf pine gradually gave way to longleaf pine. In this chapter, we will focus on oak and oak–shortleaf communities.

Historically, the composition of the understory varied, but generally was dominated by grasses. In some cases, forbs (broadleaf herbaceous plants) were dominant. Bluestems (Figure 15.2), such as big and little bluestems along with broomsedge, likely were the dominant grasses, but other common grasses would have included Indiangrass, povertygrass, silver plumegrass, and splitbeard bluestem.

Forb communities could be quite diverse (Figure 15.3a) and may have included legumes, such as beggar's-lice, partridge pea, roundhead and slender lespedeza, Illinois bundleflower, and sensitive briar (Figure 15.3b). Other common forbs included the coneflower, woodland sunflower, goldenrod, beebalm, and blazingstar (Figure 15.3c). In most cases, more fire-tolerant hardwood species, such as oaks, hickories, and blackgum, would have persisted as woody sprouts in the understory—unless fire was intense and frequent enough to have precluded tree species. Shrubs and brambles, such as blueberries and blackberries, also would have been common.

Regardless of past composition, the absence of a midstory is another important characteristic of savannahs and, to a lesser extent, woodlands. Shrubs and trees that would normally grow into the midstory either have been top-killed or eliminated, depending on fire intensity and frequency (Figure 15.4). On fire shadows (areas protected from burning by some physical feature, such as rock ledges, drains, coves, and the areas on the downwind side of those natural fire barriers), moister sites, and in many woodlands, there may be a sparse midstory, but it is much less developed than what we see in today's hardwood forests.

What has become of woodlands and savannahs?

Woodlands and savannahs were common across the Piedmont, the Great Valley of the Appalachians and surrounding smaller valleys and ridges, the Cumberland Plateau (especially the western portion), the Bluegrass, the Nashville Basin, the Great Barrens, the perimeter of the Prairie Peninsula, the Ozarks, and the Ouachitas (Figure 15.5). However, they virtually have disappeared from today's landscape. Because savannahs had few trees, they were prime areas for early settlers to clear for crops. Settlement also led to fire suppression, a trend that peaked during the late nineteenth and early twentieth centuries (see Chapter 9), thus allowing areas that had not been cleared to revert to woodland or forest conditions. Development of roads and open crop fields further reduced the spread of fire.

Today, ecologists consider these communities among the most imperiled in North America, with less than 1% of their original area remaining.

(a)

(b)

Figure 15.2 Common understory grasses associated with woodlands and savannahs include big (a) and little (b) bluestems. These grasses thrive with fire, providing a good fuel source for maintenance burns, good wildlife nesting cover, and even good forage for cattle and, historically, bison and elk. (a: Photo courtesy of C. Coffey; b: Photo courtesy of A. Vander Yacht.)

(a)

(b)

Figure 15.3 Forbs associated with woodlands and savannahs can contribute to diverse communities (a) and include legumes such as sensitive briar (b). *(Continued)*

(c)

Figure 15.3 (Continued) (c) Showy species, such as blazing star. (Photo courtesy of A. Vander Yacht.)

Those that remain often are small, fragmented remnants that were not converted because they occur on shallow soils or other areas not suitable for agriculture. These remnants often are in poor condition because fire regimes were disrupted, resulting in less frequent, more intense fires. In many cases, nonnative species crowded out native species, changed fuel loads, and/or further altered fire regimes. Even native species that are not fire tolerant may have served to degrade remnant sites by contributing to "mesophication" (see Chapter 2).

What are some of the benefits of woodlands and savannahs?

Take a walk in a savannah, and you will understand the aesthetic appeal of these open, park-like communities. They are also productive habitat for many wildlife species. The ground layer and associated fast-growing, early successional vegetation captures most of the site's productivity and provides ample cover and abundant food for numerous wildlife species. These communities have become scarce in the face of modern forestry and agriculture, but they are important to species such as northern bobwhite, Bachman's sparrow, eastern cottontail, and white-tailed deer. Species that are declining, such as the red-headed woodpecker and a number of bat

(a)

(b)

Figure 15.4 Woody midstories can develop rapidly (a), especially in regions where rainfall is ample (>40 inches per year), and preclude development of an herbaceous layer. Fire can top-kill woody stems allowing for the development of grasses and forbs (b). (a: Photo courtesy of C. Harper; b: Photo courtesy of P. Keyser.)

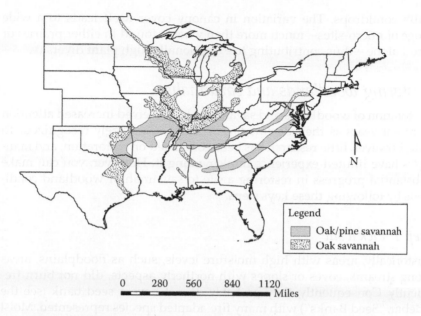

Figure 15.5 Approximate distribution of presettlement oak and oak/pine savannahs in central North America. (Adapted from Haney, A. and Apfelbaum, S.I., Characterization of midwestern oak savannahs, in F. Stearns and K. Holland, eds., *Proceedings of Midwest Oak Savannah Conferences*, U.S. Environmental Protection Agency, Chicago, IL, 1993; Henderson, R.A. and Epstein, E.J. Oak savannahs in Wisconsin, in LaRoe, E.T., Farris, G.S., Puckett, C.E., Doran, P.D., and Mac, M.J., *Our Living Resources: A Report to the Nation on the Distribution, Abundance, and Health of U.S. Plants, Animals, and Ecosystems*, U.S. Department of the Interior National Biological Service, Washington, DC, 1995.)

species (i.e., the eastern red bat, big brown bat, and northern long-eared bat), which require or prefer open midstories to forage for their insect prey, also benefit from woodlands and savannahs.

That part of the site's productivity not captured by the vigorously growing ground layer is invested in the overstory. Because of their wide spacing, oaks and other fire-tolerant species are able to develop full crowns and have the potential to produce plentiful mast—a valuable food source important to scores of wildlife species (see Chapter 4). The overstory in woodlands also provides valuable structure for wild turkey, eastern gray and fox squirrel, and forest birds that nest and forage in the canopy. The endangered red-cockaded woodpecker once used cavities in mature shortleaf pines for nesting and the open understory for foraging. Many rare plants, long absent from eastern oak communities, also respond to increased light and scarification from periodic fires. Examples include purple milkweed, Atlantic camas, slender lespedeza, plain gentian, and

Bull's coraldrops. The variation in canopy cover itself leads to a wide range of micro-sites—much more than what is found in either prairies or forest alone—thus contributing to exceptionally high plant diversity.

Restoring woodlands and savannahs

Restoration of woodlands and savannahs has received increased attention in recent years as their decline has become more widely recognized. To date, however, little research has been conducted on restoration, and managers have limited experience implementing it. However, you can make substantial progress in restoring a site to savannah or woodland conditions by following these keys steps.

Step 1: Pick a site

Historically, areas with high moisture levels, such as floodplains, areas along streams, coves or slopes with northerly aspects, did not burn frequently. Consequently, they probably do not have a seed bank (see the sidebar, "Seed Banks") with many fire-adapted species represented. Moist areas also are more difficult to burn. Other areas that are difficult to burn include small tracts with irregular boundaries and those that are steep. Such areas have a high risk of fire escape or smoke problems and should be avoided. (See Chapter 9 for more details on prescribed fire.) In areas where fire is difficult or risky to apply, it is less likely to be used and, in turn, restoration will be less successful. Conversely, areas that are easy to burn because of natural firebreaks, such as streams, wetlands, roads, or fields, are good candidates for restoration sites.

SEED BANKS

"Seed bank" is a term that refers to the viable seed stored within the top few inches of the soil. Such a seed typically is dormant, but given the right conditions (i.e., soil scarification, appropriate temperature and moisture) can germinate. Depending on past land use, seed banks can be rich in native species important to savannah restoration. With a long history of intensive agriculture or excessive erosion, seed banks are often depleted. Seed banks can also harbor undesirable species that can make savannah restoration difficult.

Also, consider site productivity. Areas capable of growing high-value timber have an increased risk of fire damage to valuable trees. Undertaking a restoration project on such productive sites should be weighed against possible economic loss. Fortunately, these are often the same sites with high moisture levels mentioned earlier. As a rule, low-quality sites do not have high-quality timber and are better suited for burning because they typically are drier.

Consider access. Timber harvest may be an important part of the restoration process and will require good haul roads. Getting equipment to the site for burning also places a premium on good access. If you construct access roads, perhaps as a part of a timber harvesting operation, try to establish them where they could later serve as a firebreak.

Stand age is also important when selecting a site. Stands that are old enough to allow for a commercial timber sale are easiest to work with because unwanted trees normally can be removed by logging. Also, remaining trees in an older stand will have thicker bark and, therefore, are less likely to be damaged by fire (Figure 15.6a). You also could consider younger stands, where most trees are large enough to survive burning (normally greater than 6 inches DBH) but too small for harvest (normally less than 12 inches DBH). However, thinning these stands is more difficult because they cannot be treated through commercial harvest (Figure 15.6b). Fire may be used to thin them, but it may take repeated fires to eliminate the larger stems. Furthermore, you cannot easily regulate the process to achieve specific results. Consequently, you may remove desirable stems, leave undesirable stems, or have excessive patchiness in your residual stand.

Consider acreage. Because shading from adjacent forests can suppress understory response, stands that exceed 10 acres are best. Exceptions include small woodlots surrounded by fields, or sites along the edge of a larger block of forest that are adjacent to a field. In either case, ample light from the field edge ensures successful understory development. If the stand will be logged, it may require a larger minimum stand size because of the practical and economic logistics of timber harvesting operations. Depending on the quality and volume of timber to be removed, this may require a unit no smaller than 25 acres unless you are treating multiple stands across the property. But a number of wildlife species associated with savannahs may not respond to units this small. On the other end of the scale, a restoration area could be as large as is practical for you to maintain with prescribed fire, perhaps a few hundred acres. For many wildlife species, larger blocks of habitat are preferable to smaller areas.

Step 2: Get rid of unwanted trees

Once an appropriate site is selected, the next step is assessing the need for timber harvest. If stocking levels are too high to allow ample light to reach the ground for either a woodland or savannah, the overstory must be thinned. Some exceptions may include stands on poor sites or sites that have had some past disturbance, such as wind or ice damage, insect

(a)

(b)

Figure 15.6 A mature, 100-year-old stand suitable for restoration (a). Note the large mature oak in the foreground. Such stems can withstand subsequent burning. Also, note the two-aged structure of this stand, indicating a formerly more open condition, or perhaps some other past disturbance, such as logging. Younger stands (b) can be suitable for restoration if there are many smaller stems that could be eliminated through burning. (Photo courtesy of A. Vander Yacht.)

outbreak (e.g., Southern pine beetle or gypsy moth), intense fire, or a recent timber harvest. As mentioned in the previous section, young forests may not need to be thinned either.

Preferably, the stand should have enough timber to support a commercial harvest. Such harvests allow for the work to be accomplished relatively quickly, selection of stems to retain, and potential net income from timber sale proceeds.

To thin young stands or those that otherwise cannot be commercially harvested, it is best to kill undesirable trees by girdling, spraying the wound with herbicide, and leaving them standing, as opposed to felling the trees and spraying the stumps. Prescribed fire can be used to thin young stands with trees less than about 4 inches in ground diameter. Standing dead trees fall apart in 3–5 years. These snags are a critical habitat component for many wildlife species. If unwanted trees are felled initially, a tremendous amount of debris accumulates on the ground in a short time and may result in excessive fuel loads. You may want to develop a woodland first if you are working with a younger stand. The higher residual stem density of woodlands can help suppress some woody vegetation while allowing the leave trees to develop more fully. At that point, additional trees can be removed, if desired, to convert the site into a savannah.

When planning a harvest, you will need to decide how many and which trees to remove. Reduce the canopy below 30 square feet of basal area per acre for savannahs and below 60 square feet for woodlands. For a mature stand on fair to good sites with leave trees that average about 14 inches DBH, this will mean leaving about 30 (savannahs) and 60 (woodlands) trees per acre. These targets will vary based on the size of your leave trees. On poorer sites, individual stems will be smaller, their canopies will be smaller, and a few more may be left—perhaps 50 (savannahs) or 80 (woodlands) per acre assuming 10–11 inches DBH for leave trees. Leaving a few extra trees per acre is a good idea because of the risk of mortality from thinning stress, wind-throw, ice, or fire. In the first 5 years after harvest, the overstory will likely become more open as some trees are lost. Regardless of specific basal area or tree per acre targets, the primary objective, where savannahs are the goal, is to open the canopy enough (70%–90% of sunlight reaching the ground) to foster vigorous herbaceous groundcover. For woodlands, 30–40% of sunlight must reach the forest floor to stimulate desirable groundcover (Figure 15.7).

Follow these simple principles when selecting residual, or "leave," trees. Because of future burning, all residual stems should be fire-tolerant species, including white, chestnut, burr, post and black oaks, and in the Lakes States, northern pin oaks. Other species that should be considered include shortleaf pine, hickories, and blackgum. Other oaks that may be

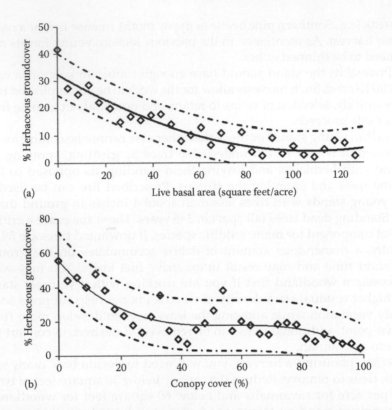

(a)   Live basal area (square feet/acre)

(b)   Conopy cover (%)

Figure 15.7 Herbaceous groundcover is strongly influenced by canopy cover. Note that herbaceous groundcover increases after the overstory is reduced (a) below 60 square feet/acre (woodland target). Further development of the herbaceous layer occurs once the canopy cover is reduced (b) below 30% (savannah target). As herbaceous cover increases, herbaceous richness and diversity increase as well.

worth retaining include Southern red, Northern red, and scarlet, but they are not as fire tolerant as the others. In fact, if the stand has a considerable component of Northern red oak, the area is probably too moist and savannah restoration is not appropriate.

Some fire-tolerant species may be vulnerable to fire because of past damage. Hollow trees can become "chimneys" when burning and may increase the chance of fire escaping. Such trees have wildlife value, but weigh retaining them against the risks. Those far from a firebreak or stand boundary can be left, but if they are close, consider them a fire hazard that will need to be raked around or removed.

Fire-intolerant species should be removed, including maples, yellow poplar, white, and Virginia pine. Many of these trees are prolific seeders and will continue to restock the ground layer with undesirable vegetation, which may become well established during intervals between fires.

Although fire typically controls such seedlings and saplings, they compete with more desirable herbaceous species, can become well established in the understory, and slow the pace of restoration.

Also, consider the trees' commercial value when deciding which ones to retain. Depending on your objectives, you may wish to remove trees of considerable value (all veneer and grade lumber) before burning to prevent damage and lost revenue. On the other hand, many low-value trees (from a commercial perspective) such as large-crowned "wolf trees" can be quite valuable to wildlife and, therefore, are good candidates for retention. In most cases, less than 30% of the stocking in a given stand contains more than 70% of its value. Although most mature oaks have thick bark and are not likely to be harmed by the low- to moderate-intensity fires used to restore and manage savannahs and woodlands, there is always risk of damage. Other, lower quality stems can provide the desired shade and mast production and will put less value at risk. If your stand has a high proportion of valuable stems, you are probably on a high-productivity site, and savannah restoration may not be appropriate—unless you are not concerned with timber value.

Consider the vigor of individual trees when deciding which to leave. Select stems likely to survive some stress and still be alive in 10 years. For shortleaf pine, retain stems with live crown ratios (the proportion of the total tree height with living branches) greater than 20%. Given the rarity of shortleaf pine today, it would be worthwhile to keep any somewhat healthy stems as a seed source. When evaluating oaks and hickories, remove trees with several dead limbs, small crowns or signs of stem damage or rot, and retain trees with vigorous, large crowns.

If you know a stand's age, consider species longevity when choosing what to leave. On medium-to-poor sites, pines, hickories, blackgum, and most oaks in the red oak group probably will not live beyond 100 years. Restoration of stands as old as 100 years where these species are dominant requires careful consideration. Oaks in the white oak group live much longer. Another consideration regarding leaving shorter-lived species is that mast production from trees approaching their physiological age limit diminishes or ceases.

Regardless of which trees you select for retention, some care should be taken to ensure spacing across the stand is not excessively irregular. Some variability is desirable, but you should not allow it to become extreme.

If you have a contractor (logger, consulting forester, etc.), conduct the harvest or herbicide treatment, be sure to have a contract that protects your interests and clearly specifies removal criteria and penalties and thresholds for damage to residual stems. Also, trees to be retained should be marked by someone representing your interests rather than those of the loggers or other contractors. Clearly establish expectations before you start so that you avoid headache and heartache later.

Step 3: Burn

Once you have opened the canopy to the right level, it is time to reintroduce fire. Fire is an absolutely critical component of restoration because it controls hardwood midstories and influences understory composition. Fire stimulates germination from the seed bank and subsequent growth of key fire-adapted species that characterize woodland and savannah understories. In fact, many herbaceous species associated with savannahs and woodlands require fire to scarify the seed for germination to occur. Furthermore, woody midstories can substantially limit the development of herbaceous vegetation in the ground layer. Much like the overstory canopy, midstory canopies can produce enough shade that regardless of the openness of the overstory, no herbaceous vegetation will grow. Fire is critical for both reducing midstories and keeping them from reemerging (Figure 15.8).

The initial burn should be conducted about 2 years after harvest to allow woody species time to germinate or re-sprout and grow. The initial burn then can begin the process of killing unwanted hardwood rootstocks. Another value of the delay in burning is that it can allow fine fuels to build-up where logging activity has broken their continuity.

Where fuel loads are high, delaying the initial burn following harvest allows them to diminish to more manageable levels. If fuel loads from logging slash or limbing/topping trees in place are excessive, you may need to delay burning for an additional year or two. Waiting more than 4 years—or even 3 years on a productive site—is not advisable because regrowth of woody stems may become too dense to allow fire to carry or they may be so large that to kill them would require fires intense enough to damage

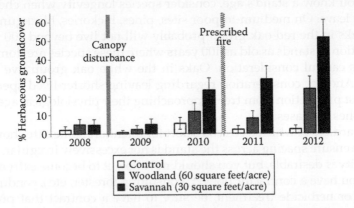

Figure 15.8 Trends in herbaceous groundcover, 2008–2012 for different management treatments during an oak woodland and savannah restoration experiment at Catoosa Wildlife Management Area, Tennessee. Overstory disturbance produced desirable gains in total herbaceous groundcover, and subsequent fire further enhanced this effect. Note that groundcover increased where canopies were more open.

leave trees. Seek assistance from experienced and qualified foresters, wild-life managers, or controlled burn contractors if you have questions.

Because growing-season burns, especially in late summer, are more effective at killing hardwood rootstocks, you should conduct the initial burn between mid-August and late September. In fact, the first two or three burns should be conducted during this period, depending on how well hardwood mid- and understory stem density is being controlled. All of these stems will not be killed, but controlling them is critical because they are resilient and easily can dominate the understory if not reduced early in the process.

Stems and rootstocks in some stands may be 80–100 years old, or more, and will not be controlled with only one or two burns. A good tar-get for woodlands and savannahs is to almost completely eliminate the midstory (less than 10 stems per acre) and reduce hardwood cover in the understory to less than 25%. Growing-season burns are helpful in pro-moting big and little bluestems, broomsedge, and other native grasses on sites where they have disappeared under forest canopies. In addition to providing important structure, these plants provide a reliable source of fine fuel for future burns.

Subsequent burns should occur within 2-year intervals, depending on the response and composition of groundcover, especially the woody response. On drier sites and regions (Ozarks, Ouachitas), fire inter-vals could be longer than on sites with greater precipitation and/or soil fertility. Also, as restoration progresses, fire-return intervals may increase as woody vegetation is suppressed. In the first few years, you should maintain fire on frequent intervals to ensure control of woody vegetation. Fire regimes for woodlands do not differ materially from those of savan-nahs. Fire in savannahs typically may be a little more intense than those in woodlands because of the increased sunlight. Canopy cover rather than fire regime is responsible for most of the difference in these two systems.

As the size and density of hardwood sprouts and saplings are reduced, you should shift some burns to the dormant season. Fires during this time of year are not as good at hardwood suppression as growing-season burns, but they help foster the development of a number of cool-season plant species important to the savannah and woodland communities, including wildryes, panic grasses, and needlegrass.

Dormant-season fire is often easier to implement because weather conditions in February and March are more favorable to burning than in late summer. After the first few burns (probably all during the grow-ing season), alternating between dormant- and growing-season burns is a good approach. In time, you may have two or more dormant-season burns for each growing-season burn. Deciding which type of burn to use should be based on conditions each year and the need to control woody growth versus fostering cool-season species.

Regardless of season, fire intensity should remain low to moderate. A good target for burn intensity is to have flame lengths between two and three feet with most of the fine fuels in the burn unit being consumed. Fires with greater intensity (i.e., flame lengths above 4 feet) may damage canopy stems and have increased risk of escape. Less intense fires may do a poor job of killing mid- and understory hardwood stems.

A word about fuel loads

Burning can be accomplished with minimal damage to mature, fire-tolerant trees, like several species of oaks and hickories. However, they are still more susceptible than shortleaf and longleaf pines. When burning where fuel loads are excessively high, such as would be typical if you felled a large number of trees during the thinning process and did not remove them, choose conditions that will not allow heavy fuels to burn. This is especially true in younger stands with smaller and less fire-tolerant stems. Ice storms or insect outbreaks that result in a large number of dead trees or broken tops accumulating on the ground are other scenarios that require caution. More moderate fuel loads still can be a problem where logging slash has accumulated around the base of residual stems. In that case, you may want to require loggers to move slash away from the base of residual trees during harvest. Also, exercise extra care where rank cover of flammable grasses, such as broomsedge, is found under hardwoods. See Chapter 9 for more details on burning and burn prescriptions.

A word about herbicides

Fire is an essential tool in restoring woodlands and savannahs. There is no suitable alternative. Herbicides, however, are a tool that may prove helpful and, in some cases, may be an essential compliment to a burning program. Selective use of herbicides can control undesirable trees that loggers would not or could not remove, or that low-intensity fire will not kill. You can use herbicides to kill midstory stems missed by fire or before a burn, where stem density precludes fire spread. If you are not able to burn as frequently as necessary, use herbicides to kill larger encroaching saplings.

Applying herbicide where there is a hardwood overstory requires specific techniques and care in selecting the appropriate herbicides to prevent damage to desirable species and trees. More detail on the use of herbicides in oak systems can be found in Chapter 10.

A word about invasive plants

Many nonnative plants can be aggressive and dominate native plant communities. Some species, such as kudzu, sericea lespedeza, and multiflora rose, are well-known and serious pests, but there are a number of others that can negatively affect restoration success. They can out-compete more desirable native species, suppress germination, degrade structure, and alter fuel loads. Roads are one of the key pathways for introducing nonnative invasive plants. (If you build any haul or access roads during the restoration process, do not sow them with any species that may be invasive—this is often where they get their start.) Areas that flood frequently often have problems with invasive weeds as well. Excessive populations of Johnsongrass, tall fescue, Bermuda grass, serecia lespedeza, Japanese stiltgrass, and miscanthus near your site may preclude successful restoration. Use herbicides to control them early on, *before* they become too abundant to control. Conversely, experience has shown that in the interior of restoration sites, growth of native groundcover from the seed bank can be thick enough to limit encroachment of many of these invasive plants.

When am I done?

Restoring an ecosystem—or at least a plant community—that has disappeared over a period of more than 100 years does not happen overnight. Well-entrenched hardwoods will not quickly relent, even in the presence of such profound disturbances as timber harvest and fire. It will take patience and persistence over 10 years or longer, depending on the site, to realize your restoration goals. Progress will come more quickly on drier sites, or on sites that were more recently in a woodland or savannah condition. More time is required to achieve satisfactory results on sites with greater fertility or available moisture and a longer absence of open conditions.

Regardless of how long it takes to get a site to resemble a woodland or savannah, keeping it there is an issue of ongoing "maintenance." After all, a woodland or savannah is a transitional community perpetuated by periodic disturbance. In longleaf pine country, landowners who have retained savannah-like conditions over most of the past century have been diligent in sustaining appropriate (i.e., 1–2-year frequency) burning regimes. We should expect no less in oak and oak/shortleaf country.

Regeneration

Like any forest community, trees retained in woodlands and savannahs eventually die. Fortunately, seedlings and saplings of desired species are relatively fire tolerant and may persist under burning for many years. However, at some point, they may decline in numbers and there might not be enough remaining to provide adequate regeneration. Careful observation of regeneration is important so that timely steps can be taken to foster regeneration as needed.

If the midstory and most advanced regeneration of desirable species has been greatly reduced, you may want to protect some oaks and/or shortleaf pine regeneration seedlings from fire until they are large enough to survive a burn. This can be accomplished with longer fire-return intervals, less intensive fire, and dormant-season fire. Allowing woody stems to reach a 3- to 4-inch ground diameter will enable them to survive fire. Less desirable woody species also will exploit the reduction in burning. Therefore, you should not protect regeneration until after the more fire-intolerant species have been eliminated or severely reduced in density. In most cases though, there will be enough oak or pine reproduction that make it through your burning regime that it will be unnecessary to suspend burning. Once the desired regeneration has grown to an adequate size, you can resume growing-season burns and increase intensity as needed.

Let's get started!

No one who has ever strolled through a savannah, whether pine or oak, will ever forget its sublime beauty. Professional landscape architects pay homage to savannahs when designing parks and the settings of great mansions, where scattered large trees provide partial sunlight to reach the grassy lawns beneath. Beauty aside, these sites offer a glimpse into the primeval conditions once common across this part of the continent. They also offer great hunting and camping sites. Most important, they make a tremendous contribution to conservation, whether for bobwhite, grassland birds, rare wildflowers, or for an all-but-lost part of our oak ecosystems. Taken all together, savannahs and woodlands are a legacy we cannot allow to disappear.

A GOOD EXAMPLE

The Big Piney Ecosystem Restoration Project on the Ozark–St. Francis National Forest demonstrates the benefits of oak and pine woodland restoration and management. The Nature Conservancy, Arkansas Natural Heritage Commission, Arkansas Forestry Commission, private landowners, and others host field visits to the demonstration area to promote the success of this collaborative project to restore the oak/hickory and pine/oak ecosystems of the Ozark Highlands on 60,000 acres of the Big Piney Ranger District.

Specific project activities include the application of periodic prescribed fire and thinning by commercial and noncommercial methods. Since its inception, prescribed fire has occurred on the entire 60,000 acres with several thousand acres of growing-season burns.

Figure 15.9 Extensive savannah and woodland restoration in the interior highlands of Arkansas has led to healthy oak–pine communities such as seen in this photograph. (Photo courtesy of P. Keyser.)

A comprehensive monitoring program documenting ecosystem responses to prescribed fire and woodland thinning treatments guide management decisions. The monitoring program includes tracking changes in overstory, understory, and herbaceous plant species, the structure of plant communities, fuel loads, nonnative invasive species, and breeding and wintering bird communities (Figure 15.9).

CATOOSA SAVANNAH RESTORATION PROJECT

An unusually severe pine beetle infestation during the mid-1990s resulted in a timber salvage operation on 1555 acres during 1999–2001. As the salvage progressed, the response by the plant community was dramatic. Native warm-season grasses, wildflowers, native legumes, and other forbs that had been dormant in the seed bank or suppressed by the overstory were stimulated by canopy removal.

To further encourage these plants, prescribed fire was initiated on 518 acres in 2002. The potential to restore a unique and imperiled community was evident. Based on this initial success, a commitment was made to restore a savannah on the site. By 2007, with help from the University of Tennessee's Center for Native Grasslands Management, Quail Unlimited

Figure 15.10 Catoosa Wildlife Management Area savannah restoration project has been successful because of a large group of partners working together cooperatively. (Photo courtesy of K. Kilmer.)

and the National Fish and Wildlife Foundation, the project had grown to 3730 acres, perhaps the largest savannah woodlands restoration project east of the Mississippi River.

The savannah has continued to develop through a combination of forest thinning and prescribed burning. The open oak canopy now in place has allowed savannah and woodlands to flourish and, with it, wildlife species dependent upon an early successional habitat. The thinner canopy allows more sunlight to reach the ground, leading to a diverse profusion of native plants that readily respond to prescribed fire.

The university evaluated the flora and wildlife response to the changed habitat conditions. The results showed that canopy reduction plays a crucial role in the restoration process. Increased light reaching the ground resulted in an increase in grass and forb cover and species richness. Although a number of the forest bird species declined following the disturbance, early successional bird species increased markedly.

The savannah continues to serve as an education tool for land managers and other interested groups in the Eastern United States (Figure 15.10).

Further suggested reading

Barrioz, S., P. Keyser, D. Buckley, D. Buehler, and C. Harper. 2013. Vegetation and avian response to Oak Savanna Restoration in the Mid-South USA. *American Midland Naturalist*. 169: 194–213.

Brock, T.D. and K.M. Brock. 2004. Oak savannah restoration: Problems and possibilities. University of Wisconsin–Madison and Savannah Oak Foundation, Madison, WI.

Burger, G. and P. Keyser. 2013. Ecology and management of oak woodlands and savannahs, PB 1812. University of Tennessee Extension, Knoxville, TN.

Davis, M.A., D.W. Peterson, P.B. Reich, M. Crozier, T. Query, E. Mitchell, J. Huntington, and P. Bazakas. 2000. Restoring savannah using fire: Impact on the breeding bird community. *Restoration Ecology*. 8: 30–40.

Engstrom, R.T. 2010. First-order fire effects on animals: Review and recommendations. *Fire Ecology*. 6(1): 115–130.

Haney, A. and S.I. Apfelbaum. 1993. Characterization of Midwestern oak savannahs. In: F. Stearns and K. Holland, eds., *Proceedings of Midwest Oak Savannah Conferences*, U.S. Environmental Protection Agency, Chicago, IL.

Henderson, R.A. and E.J. Epstein. 1995. Oak savannahs in Wisconsin. In: E.T. LaRoe, G.S. Farris, C.E. Puckett, P.D. Doran, and M.J. Mac. *Our Living Resources: A Report to the Nation on the Distribution, Abundance, and Health of U.S. Plants, Animals, and Ecosystems*. Washington, DC: U.S. Department of the Interior National Biological Service.

Peterson, D.W. and P.B. Reich. 2001. Prescribed fire in oak savannah: Fire frequency effects on stand structure and dynamics. *Ecological Applications*. 11: 914–927.

Siemann, E., J. Haarstad, and D. Tilman. 1997. Short-term and long-term effects of burning on oak savannah arthropods. *American Midland Naturalist*. 137(2): 349–361.

Wilgers, D.J. and E.A. Horne. 2006. Effects of different burn regimes on tallgrass prairie herpetofaunal species diversity and community composition in the Flint Hills, KS. *Journal of Herpetology*. 40(1): 73–84.

chapter sixteen

Managing bottomland oaks

Mike Staten and Wayne K. Clatterbuck

Contents

Ecology of bottomland oaks 248
Species/site relationships 248
Stand development 250
Management emulating natural disturbance regimes 251
Considering the landscape 251
Forest structure and diversity 252
Managing animal competition 252
Other management considerations 253
Silviculture: Making the rotation 253
    Mature forests, perpetuating oaks in the stand 253
    Regeneration: Producing young forests 254
    Mid-aged forests: Producing mature forests 255
    Thinning guidelines 256
    Young forests: Producing mid-aged forests 257
Crop tree release 257
Bole quality 257
Acorn production 258
Conclusion 258
Further suggested reading.. 258

The major river systems of North America were once covered with bottomland hardwood forests. Because of the fertility of these alluvial soils, many of these forests were converted to agricultural land. For example, the alluvial valley of the Mississippi River once contained more than 24 million acres of bottomland hardwood forests! Even with restoration of hardwood forests through the Wetland Reserve Program (WRP) and the Conservation Reserve Program (CRP), less than 5 million acres remain in forests today.

These forests provide essential ecosystem functions and values, including capturing sediment, slowing water velocity, and providing wildlife habitat. Ecosystem values refer to the benefits derived from the functions performed, such as aesthetics, ecological services, and recreation.

Ecology of bottomland oaks

In bottomland forests, the period of time a site is subjected to standing water and saturated soil is the major determinant of which oak species occur. For example, the overcup oak is considered the most adapted to sustained periods of standing water, whereas water oak and cherrybark oak are found on sites with shorter durations of saturated soil. Soil pH also plays an important role in determining which oak species colonize a particular site. Soils closest to major river channels often have a higher pH (more basic, less acidic) and, in most cases, are conducive for establishment and growth of riverfront hardwoods, such as cottonwood, sugarberry, and sweet pecan. As soils age and mature, pH begins to drop to more acidic levels and oaks, which are more competitive on such soils, gradually become a more prominent component of the stand.

Nutrient input is an important consideration for any landscape. In the case of bottomland ecosystems, nutrients deposited from soils eroded upstream drive the system. For example, sediment and nutrient loads that come downstream to the bottomlands along the Mississippi River in Mississippi and Arkansas are originally eroded from fertile landscapes of the prairie states in the Midwest. Conversely, nutrient loads originating from sediments from poorer, less fertile landscapes may not be as productive. A good example of this is the numerous smaller rivers along the Coastal Plain that rise in the less fertile sandy ridges of the region. They are often referred to as *blackwater rivers* because their waters are characteristically dark, stained by tannic acid from decaying organic material. Because of the sandy landscapes from which they originate, they lack nutrient-rich silt or clay particles and, without these eroded sediments, the water is not muddy. Nutrients in blackwater river systems are replenished by the deposition of organic matter rather than silt.

High soil fertility may hinder the ability of oak to capture a site because other species can establish a root system more quickly, grow rapidly, and outcompete many oak species. Once such competition has become well established, it can easily overtop the more shade-intolerant oak reproduction. Conversely, low soil fertility may favor oaks by allowing them time to establish deep root systems, something they do at the expense of height growth during the early stages of development.

Species/site relationships

First or major bottoms are the active floodplain of major drainage systems (see Figure 16.1). Over hundreds of years, river channels move, leaving a ridge and swale topography, which over time will continue to receive sediment deposits. Since the heavier sands drop out quickly

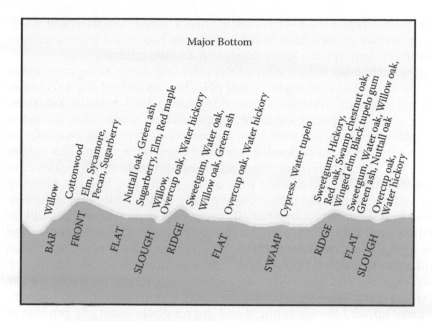

Figure 16.1 Topographic variations and associated species in a major stream valley. (Figure courtesy of John D. Hodges.)

near the channel, they create fronts or natural levees, while silt and clay particles are suspended in the water longer and drop out farther from the river where there is little or no stream flow. The swales may become shallow flats through sediment deposition, whereas deeper swales or sloughs that hold water longer are referred to as *backswamps*. The term backswamp derives from the fact that on most floodplains the lowest elevations are found furthest from the river front (near the bluff or uplands), hence they are at the back of the floodplain. The ridges were once natural levees that now are well removed from the front due to natural shifts in the course of the stream and have been covered by layers of silt. Because of the underlying sand deposits, they are well drained, but because of the silt, they are typically quite fertile and are among the most productive sites within a floodplain.

Minor floodplains are tributaries that flow into major floodplains. Soils in minor floodplains are normally more acidic than major floodplains. Unless there is heavy erosion upstream, sedimentation is reduced and stream channel changes are less prevalent. Undulation of ridges and swales is shallower either because of less stream channel change over time or sediment has filled the deeper swales. Similar species/site relationships occur in minor floodplains as in major floodplains. Both major and minor floodplains contain a first bottom (the active floodplain) and usually a second bottom, or terrace. Terraces are former active floodplains that existed before the river system eroded a deeper and narrower flood-plain. Terraces, which flood only in extremely high-water events, serve as transitions between the active floodplain and the uplands.

As mentioned before, alluvial deposits laid down along most major bottoms are basic with regard to soil pH. The species best able to colonize natural levees are riverfront hardwood species, which do not include oaks. Since most oak species are more competitive on acidic soils, the aging or maturing of the soil over time creates conditions more favorable for them. Oaks can become dominant on the ridges, terraces, and flats farther from the river. Cherrybark, Shumard, swamp chestnut, and water oaks will colonize the well-drained soils on the ridges and terraces. Overcup oak will succeed in the wetter, more poorly drained flats, and sediment-filled backswamps that traditionally grow bald cypress and black willow. Willow and Nuttall oaks will develop on the transitional areas between the terraces and backswamps depending on the amount of sedimentation and water saturation. Competition from sweetgum, boxelder, sugarberry, maples, elms, and green ash may cause establishment problems for oak seedlings, but these same competitors play a major role in training oaks to grow upward toward sunlight and improve bole quality by pruning or rubbing limbs from the established oak saplings and poles.

Stand development

Stand development involves changes in stand structure over time. The following explanation of these dynamics is heavily based on that provided by Dr. Chad Oliver (see his excellent treatment of this subject listed under, "Further Suggested Reading"). These changes occur horizontally (distribution of trees by species and diameter) and vertically (distribution of tree heights) within each stand. These changes typically occur in four stages: stand initiation, stem exclusion, understory initiation, and old growth. Recognizing these stages will help practitioners prescribe management practices necessary to promote oak in the forest.

The first stage, stand initiation, is the capture of an opening created by a disturbance by tree reproduction vying for available sunlight, moisture, and nutrients along with many other plants. The second stage, stem exclusion, begins when individual stems of the most competitive trees shade out less competitive stems. Once the surviving, dominant stems reach a height where sunlight can penetrate the canopy of tree crowns down to the forest floor; the third stage begins when the shade-tolerant species initiate an understory layer. The final stage, old growth, will become more complex and diverse as vegetation layers are added to the stand that create both vertical and horizontal stratification. Though valuable for forest stand diversity, understory and midstory layers of shade-tolerant species will become a nemesis for oak reproduction in the future.

Management emulating natural disturbance regimes

Oaks are adapted to disturbance regimes, such as fire, wind, ice, grazing, and flooding. Silvicultural disturbances designed to establish and maintain oak species in a forested context should emulate, as closely as possible, these same natural disturbances (see Chapters 5, 7, and 9).

Considering the landscape

Streamside management zones (SMZ) are unharvested or lightly harvested buffers foresters or land managers establish along perennial, intermittent, and sometimes, ephemeral streams. Their purpose is to intercept surface water runoff and, therefore, reduce the amount of sediment transported into the stream. State forestry agencies have established best management practices (BMPs) stipulating minimum widths for buffers between the stream and the managed stand. Because all bottomland stands are near a stream, it is important to be aware of guidelines and requirements of BMPs in your state. Refer to your local state BMP guidelines for additional information.

Corridors of unmanaged or lightly managed forest connecting areas of special concern with similar areas of unmanaged or lightly managed forest can add habitat diversity to the tract. In addition to providing diversity, corridors often facilitate travel for various wildlife species by connecting SMZs to one another. Corridors may play an important role by enabling smaller, less mobile species to move across landscapes that otherwise may not be suitable for them and thus maintain genetic connectivity and viable populations. For instance, pond-breeding amphibians can use corridors for moving from breeding areas to wintering areas.

Research has demonstrated that landscapes managed under even-aged and uneven-aged systems with uncut reserves are more diverse than those managed under a single system. For instance, intensively managed stands in conjunction with less intensively managed stands separated by unmanaged corridors and streamside management zones create a landscape with considerable diversity. Timing of timber harvest cycles also plays a role in maintaining diversity. Light cuts on short entry intervals may produce more pleasant aesthetics, but the amount of sunlight reaching the forest floor will be short-lived as the tree crowns quickly fill canopy gaps. Conversely, heavier cuts may be less pleasant aesthetically, but allow more sunlight to reach the forest floor for longer periods of time. This additional sunlight allows oak reproduction more time to develop and become competitive before the canopy recloses and captures the sunlight. A mixture of both harvest strategies will create more diversity than either alone. Implementation of frequent light harvests, however, is dependent upon the availability of logging contractors and a market for the forest products within reasonable proximity to the property. Regardless, harvest intensity should be based on landowner goals and whether desired wildlife species require early-, mid-, or late-successional stages.

Forest structure and diversity

A diverse forest with overstory, understory, and midstory trees as well as snags, cavities, coarse woody debris, and canopy gaps provides habitat for a wide variety of wildlife species. To find additional information on components of forest structure important for wildlife, visit the Lower Mississippi Valley Bird-Habitat Joint Venture website (www.lmvjv.org/bookshelf.htm). The documents available at this site describe desired forest conditions pertinent to a diverse bottomland hardwood forest managed for wildlife. Stands of multiple age classes create the opportunity for a shifting mosaic of stands, where, as one stand becomes older, a mature one may be shifted back to a younger age through management or natural disturbance. Management can also shorten the time of progression of stand development by mimicking natural disturbances that may or may not otherwise occur within the desired timeframe. Management can also help the forest function in a manner more similar to an older, more complex stand through the creation of additional canopy gaps, allowing for development of a more layered vegetative structure.

The harvest rotation of stands is important for maintaining diversity and to perpetuate the forest with a reasonable component of oak. Rotation length is less important in uneven-aged management (i.e., single-tree selection or group selection) where the cutting interval drives the age structure. Rotation length is more important in even-aged management (i.e., shelterwood and clearcut) for maintaining age structure of an oak forest.

Managing animal competition

Feral hogs are voracious feeders that compete with native wildlife for acorns. Because of their ability to produce abundant offspring, feral hog populations can grow quickly and, as a result, limit the availability of acorns required for successful oak regeneration. Their ability to use their noses as shovels when rooting can disturb and destroy established oak seedlings. White-tailed deer may browse oak seedlings. An overabundant deer population can limit the survival and/or reduce the size and vigor of advance regeneration, thus precluding future stands of oaks (see Chapter 17). Beavers can extend the hydroperiod and ultimately change oak species composition. Oaks are normally not preferred food for beavers, but under prolonged inundation caused by beaver dams, oak mortality can occur.

Other management considerations

Green tree reservoirs are shallow, seasonally flooded impoundments built in bottomland hardwood forests dominated by oak species to create habitat for migratory waterfowl. These reservoirs can change the hydroperiod and ultimately change tree species composition. Although designed to hold water for attracting ducks, it is imperative water be properly controlled. Inundation before dormancy and after bud break will adversely affect standing oak trees and lead to mortality or even loss of the entire stand over time.

Bottomland hardwood forests can be affected by invasive species, such as tallow tree, Japanese honeysuckle, Chinaberry, and privets. In forested wetlands, tallow trees can be extremely invasive and outcompete oaks and other native species in canopy gaps. The tallow tree competes well in bottomland hardwood areas from the Gulf of Mexico northward for about 200 miles, but is killed or damaged by extended periods of freezing temperatures further north. It is best to remove these invasive species from your property as soon as they are detected to reduce their spread and colonization, but most are well established throughout the Southeast.

Poor vigor makes trees more susceptible to insect and disease problems. Most insects and diseases found in bottomland hardwood forests are a normal part of the natural environment. The best way to handle these problems is by maintaining a healthy forest that is resilient to large-scale infestation or infection caused by insects and diseases. Active forest management can help maintain stand and tree vigor by removing stressed or unproductive stems and thus allow more thrifty ones room to grow. Some wildlife species, such as woodpeckers, are attracted to dying trees affected by insect and disease infection. Where wildlife is a primary goal, these natural occurrences may be allowed to run their course, with the understanding that oak reproduction may become a priority at a later date.

Silviculture: Making the rotation

Mature forests, perpetuating oaks in the stand

An old-growth bottomland oak forest would normally be composed of multiple tree species, including those that are shade-tolerant and capable of dominating the site once the oaks begin to phase out. These species may have established themselves with the initial regeneration of the stand or may have become established in canopy gaps created by subsequent disturbances. This combination of tall trees and numerous understory layers provides habitat for a multitude of wildlife species, including migratory waterfowl and songbirds, tree squirrels, wild turkeys, and black bear. Many of these species eat acorns that mature oaks produce. Regardless of the aesthetic, ecological, or wildlife benefits of a mature oak forest, it eventually will begin to change as old or unhealthy individual trees senesce and die. Dead and dying trees have value for some wildlife as den trees or coarse woody debris. Such debris produces food and cover for invertebrates that, in turn, form the base of the food chain upon which many other species subsist. For instance, many small mammals, amphibians, and reptiles feed on the invertebrates associated with woody debris; some also use it for escape cover, thermal cover, and a substrate for laying eggs. But the oaks must be replaced if the forest is to persist as an oak forest.

Sunlight must reach the forest floor for oak seedlings and saplings to survive (see Chapters 5 and 6). Oaks require some sort of disturbance—flooding, fire, ice, windstorm, timber harvest—to create these light conditions. The ability to re-sprout (coppice) from established root systems also gives oaks a distinct advantage over other tree species following disturbance. Without disturbance, shade-tolerant species, such as the boxelder, sugarberry, elms, and, in minor bottoms, the American beech, will ultimately dominate the canopy and, without adequate sunlight, only shade-tolerant species will be able to continue to regenerate. This is referred to by some ecologists as a *climax forest*; oaks do not do well in the climax forests of bottomlands. It is imperative to plan for oak regeneration to perpetuate an oak component in bottomland forests.

Regeneration: Producing young forests

Two management systems are commonly used to regenerate mature bottomland oak forests: even-aged and uneven-aged (see Chapter 7). The even-aged system is designed to create and maintain numerous larger canopy gaps in which oak seedlings can survive and grow. This system is also used to release established advance reproduction, which resulted from previous canopy disturbances—man-made or natural. A clearcut may not provide adequate regeneration of oaks and furthermore, the removal of all trees in one cut may limit the seed source and, therefore, the success of a second attempt. In either management system, the key to regeneration is to start with a partial cut. In an even-aged system, use thinnings and small shelterwood or clearcut harvests. The even-aged system also creates the opportunity to divide a property into stands or cutting units to maintain a shifting mosaic of stands that vary in age, size, and area. The harvest entries associated with a final stand harvest can also be used as an opportunity to conduct the partial harvests in adjacent or nearby stands, which are needed to foster proper regeneration or achieve other objectives. Normal rotation lengths are 70–100 years, with room for shorter or longer rotations depending upon landowner goals and objectives. A shorter rotation may be desirable for timber production and for wildlife species such as deer, black bear, rabbits, and woodcock that use young bottomland forests. A longer rotation may be desired where aesthetics and wildlife that use older forests are more valuable to the landowner than maximizing timber production.

The uneven-aged system is also designed to create canopy gaps in which oak seedlings can survive and grow, albeit much smaller than those created in even-aged systems. In an uneven-aged system, group- and single-tree selection cuts (which should include removing competing species) are used to ensure adequate sunlight reaches the forest floor to establish oak reproduction. This process may require two to three entries to establish and then release established reproduction. Releasing advance regeneration is also important in an uneven-aged system. The inherent problem with the uneven-aged system is the amount of sunlight reaching the forest floor over time. For oak reproduction to establish and grow, care must be taken to make gaps large enough to maintain sunlight for an adequate period of time or, at more frequent intervals, return to the stand with treatments. The single-tree selection method requires a very good understanding of natural processes and is best performed by an experienced forester. Returning often for additional management is important. The group selection method is closely related to the even-aged system, except smaller, scattered harvest units (0.5–2 acres each) are cut within the stand or other cutting unit. The rotation length is less important, but harvest entry cycles should be frequent (7–15 years) and a cutting unit may receive thinning, single-tree selection, and group selection in the same entry event.

Mid-aged forests: Producing mature forests

Mid-aged bottomland oak forests are those at the end of the stem exclusion stage through the understory initiation stage of forest stand development. It is the transition between young and mature forests. In rough terms, this will occur after about age 20–25 or when stems on moderate sites reach about 5 inches DBH or more. After about age 50–60 years, depending on site quality, the stand will begin to take on more mature forest characteristics and transition into the mature forest stage. Managing the mid-aged forest through silvicultural manipulation may help reduce the amount of time required to develop the canopy gaps and associated understory layers characteristic of mature forests. Although these forests are not as valuable for wildlife as either mature or young forests, mid-aged forests are not devoid of wildlife. Mid-aged stands provide habitat for species such as the Acadian flycatcher, which prefers an open understory created by a shaded canopy to sally for insects. In the late winter and early spring, wild turkeys preparing for nesting will turn over leaf litter to forage for protein-rich insects in the open understories typical of these stands. Establishment of an understory through silvicultural interventions will create ground- and shrub-layer vegetation necessary for escape, feeding, and nesting cover for many wildlife species.

Thinning guidelines

Before thinning bottomland hardwood forests, there are several important guidelines to consider. First, begin thinning early in the life of the stand. Whenever economically possible, thin the stand before it becomes overstocked and possibly stagnant. Second, favor the largest trees with well-developed crowns and high-quality boles. Trees with these characteristics will likely be classified as preferred growing stock trees and should be favored when marking a stand for partial cutting. Thinning from below or low thinning (removing stems from the lower portion of a canopy first, and then working your way up into the canopy to remove additional stems until the desired residual density is achieved) is the preferred approach and is designed to remove trees with suppressed or inferior crowns, likely classified as cutting stock trees (see Chapter 11). However, even when thinning from below, do not hesitate to remove larger trees that are over-mature, damaged, or of a less-desirable species, likely classified as reserve stock trees that could be removed if they compete with a preferred growing stock trees or are at risk of dying or deteriorating before the next cutting entry. Use frequent, light thinning entries instead of infrequent, heavy thinning entries. Frequency and intensity of thinning involve trade-offs between individual tree growth (infrequent, heavy) and stand growth (frequent, light), logging damage and production of epicormic branches. Whenever a thinning is conducted, take care to avoid excessive logging damage to residual trees, which are usually crop trees. Some logging damage is inevitable (5% or so) following any partial cutting, but it can be minimized through carefully planned and supervised logging practices. Typically, any damage removing bark on an area of 100 square inches or more will lead to degradation of the stem's quality through introduction of rot or pathogens. When thinning, leave large saplings or small pole-sized stems near residual (crop) trees whenever possible to provide shade on their lower boles to reduce the risk of epicormic branching. The presence of small poles and large saplings near residual trees also will serve to protect those trees from logging damage caused by passing skidders.

Young forests: Producing mid-aged forests

Young oak forests are those at the stand initiation stage through the stem exclusion stage of forest stand development. Such forests are important habitat for the multitude of wildlife species that require more dense forest structure. In the first few years following establishment, these stands contain heavy groundcover, including young trees, brambles, vines, forbs, and grasses. Many species of migratory and resident songbirds, small mammals, rabbits, black bear, and white-tailed deer all use young forests.

A key component of oak regeneration is managing plant competition. Oaks are intermediate with respect to shade tolerance and often unable to compete with faster growing pioneer species, such as sycamore, cottonwood, persimmon, and sweetgum. Although competition from species such as sweetgum is important for pushing height growth in oak reproduction, thereby creating a long bole and efficient crown, competitive species such as sycamore and cottonwood could keep oak reproduction from becoming established or reaching the canopy. Maintaining the oak component may require early intervention through precommercial thinning, biomass thinning, or herbicide treatments that free emerging oak saplings and poles from competition from adjacent crowns of less desirable species.

Crop tree release

Crop tree release is a valuable practice that should be implemented in pole-sized bottomland hardwoods. The practice can accelerate the growth of desirable stems with good potential for timber production. It can also be used to accelerate crown development for increased mast production. For a more in-depth explanation of crop tree release, see Chapter 11. The crop tree release treatment also may be accomplished by using herbicides (also see Chapter 10). Typically, a hatchet is used to open a wound to the cambium layer inside the bark and a suitable herbicide, such as triclopyr, is squirted into the frill. Alternatively, trees can be girdled with a small chainsaw and the wound sprayed with the herbicide mixture.

Bole quality

Quality of the tree bole for future sawlogs or veneer is a result of competition of oaks with other trees. Often, the best competition comes from tree species other than oaks. There appears to be a relationship between oaks and sweetgum, possibly because of their local association on certain soil types where sweetgums serve as competition for sunlight on limbs of the lower bole of the oak trees. As the treetop is forced to grow upward for sunlight, the shaded, unproductive lower limbs die. During heavy winds, the swaying action of the intertwined sweetgum limbs and boles "scrub" the dead limbs from the bole of the oaks. Other tree species in close proximity to oaks also will perform the same function.

Acorn production

Quality of bole may not necessarily be your primary objective for all of your forested land. Acorn production, which is positively related to crown size, also may be an important goal. Early thinning releases crowns of established oaks. More open-grown oaks tend to produce acorns at younger ages, but herein is a trade-off for future bole quality. Depending on your objectives, it may be wise to choose areas where oaks are already limby to focus on this objective and save other areas with good bole quality for producing timber products.

Conclusion

Bottomland hardwood forests are characteristically diverse, highly productive, critical for production of forest products, and important habitat for numerous wildlife species. The goals and objectives of the landowner should drive the management system used to establish and/or maintain an oak component within the forest. Oaks are adapted to disturbance regimes that allow sunlight to reach the forest floor. They survive by developing a well-established root system, which enables them to vigorously re-sprout following disturbance. Under natural conditions, these disturbances include wind and ice storms, and wildfire. Management strategies employing techniques mimicking natural disturbances are critical and help achieve desired objectives in a timely manner, helping to assure enjoyment of this exceptional type of forest for years to come.

Further suggested reading

Annand, E.M. and F.R. Thompson III. 1997. Forest bird response to regeneration practices in central hardwood forests. *The Journal of Wildlife Management* 61(1): 159–171.

Brinson, M.M. and R.D. Rheinhardt. 1998. Wetland functions and relations to societal values, in M.G. Messina and W.H. Connor (eds.), *Southern Forested Wetlands: Ecology and Management*, pp. 29–48. Lewis Publishers, Boca Raton, FL, 616 pp.

Franklin, J.F., R.J. Mitchell, and B.J. Palik. 2007. Natural disturbance and stand development principles for ecological forestry U.S. Forest Service General Technical Report NRS-19, Newtown Square, PA, 44 pp.

Johnson, R.L. 1981. Wetland silvicultural systems, in *Proceedings of the 30th Annual Forestry Symposium*. Louisiana State University Press, Baton Rouge, LA, pp. 63–79.

LMVJV Forest Resource Conservation Working Group. 2007. Restoration, management, and monitoring of forest resources in the Mississippi Alluvial Valley: Recommendations for enhancing wildlife habitat. R. Wilson, K. Ribbeck, S. King, and D. Twedt (eds.), www.lmvjv.org, accessed August 31, 2015.

Lockhart, B.R., J. DeMatteis, L.A. Harris, and A. Ezell (compilers). 2005. Mississippi hardwood notes: Designed for the professional forest resource manager. Mississippi Forestry Commission, Jackson, MS [CD format].

Lockhart, R.L., A.W. Ezell, J.D. Hodges, and W.K. Clatterbuck. 2006. Using natural stand development patterns in artificial mixtures: A case study with cherrybark oak and sweetgum in east-central Mississippi, USA. *Forest Ecology and Management* 222: 202–210.

Meadows, J.S. 1996. Thinning guidelines for southern bottomland hardwood forests, in K.M. Flynn (eds.) *Proceedings of the Southern Forested Wetlands Ecology and Management Conference*, Consortium for Research on Southern Forested Wetlands, Clemson University, SC, pp. 98–101, 332 pp.

Oliver, C.D. 1981. Forest development in North America following major disturbances. *Forest Ecology and Management* 3: 153–168.

Oliver, C.D. and B.C. Larson. 1996. *Forest Stand Dynamics*, updated edition. John Wiley and Sons, New York, 520 pp.

Putnam, J.A., G.M. Furnival, and J.S. McKnight. 1960. *Management and Inventory of Southern Hardwoods, USDA Forest Service Agricultural Handbook No. 1812.* USDA Forest Service, Washington, DC, 102 pp.

Smith, R.D, A. Ammann, C. Bartoldus, and M.M. Brinson. 1995. An approach for assessing wetland functions using hydro geomorphic classification, reference wetlands and functional indices. Technical Report WRP-DE-9, U.S. Army Corp of Engineers Waterways Experiment Station, Vicksburg, MS.

chapter seventeen

Managing deer impacts on oak forests

David deCalesta, Roger Latham, and Kip Adams

Contents

Deer in the eastern oak forests prior to and soon after
European settlement.. 262
Understanding density and carrying capacity 263
Carrying capacity for diversity 265
Carrying capacity for tree regeneration... 265
Carrying capacity for maximum sustained yield of deer 266
Crash... 266
Is there a problem?... 267
Is it deer? 269
Managing deer impact 270
Reducing deer density.. 270
Protecting understory vegetation 273
Counteracting effects of prolonged high deer density on
understory vegetation 274
Selecting a management strategy 275
Monitoring: Have you solved the problem?.. 276
Further suggested reading.. 276

White-tailed deer benefit from the food and cover oaks provide. However, deer can be a detriment to a sustainable and healthy oak forest. Deer consume vast quantities of acorns, and they eat leaves and twigs of oak seedlings and saplings, acquiring fats, carbohydrates, and proteins for growth and body maintenance. At low or moderate deer densities, deer foraging is not an issue. However, too many deer can consume too many acorns and oak seedlings and may damage too many oak saplings, resulting in the elimination of or severe reduction in oak regeneration.

Deer have the potential to double their population size every 2 years. Like all wildlife, deer abundance is controlled by outside forces—food supply, weather extremes, predation, parasites, and diseases. Mortality factors can cause profound effects on deer density, which impact oak forests.

Deer in the eastern oak forests prior to and soon after European settlement

Prior to the arrival of European settlers, there was a full complement of predators—mountain lions, bobcats, red and gray wolves, and bears. These predators and the Native Americans hunted deer every day of the year. There were no seasons or bag limits, and deer of all ages and sexes were killed and eaten. Generally, predation (including hunting) cancelled out recruitment and tended to maintain a fairly constant and relatively low deer density. Tree seedlings germinated and proliferated where the forest overstory was opened by fire, wind, or ice, which occurred sporadically and randomly over the landscape. Relatively low deer density and natural disturbances helped maintain healthy oak ecosystems.

As Europeans settled in North America, drastic changes occurred in the landscape. Native Americans were nearly eliminated by European diseases, forced relocation, and warfare, while wolves and mountain lions were extirpated in most of Eastern North America to protect domestic livestock. European development of North America opened vast tracts of intact forestlands, providing increased browse and other foods as crops were planted following clearing. Deer abundance should have skyrocketed.

However, during this period, deer were nearly exterminated by market hunting in many areas of North America where oaks were part of the ecosystem. Resultant public outcries against the excesses of market hunting and the emergence of the conservation movement at the end of the nineteenth century spawned state and federal wildlife agencies and ensuing restrictions on deer hunting. The alignment in the early 1900s of heavy restrictions on deer harvest (bag limits, shortened hunting seasons, "buck only" laws), near elimination of predators, and habitat changes (timber harvest, land clearing and burning, followed by agriculture) created vast amounts of deer food, producing an explosion in deer numbers during the 1920s and 1930s across the East and Upper Great Lakes regions, including areas dominated by oak forests.

Rapid increases in deer density during the decades following predator removal resulted in the elimination of many of the plants, which deer preferred as forage from the understory, including seedlings, shrubs, and herbaceous plants. Forest understory species have notoriously short-lived seeds and many disperse only at short distances. Where deer numbers have been high for decades, the seeds bank (viable seeds in the soil) no longer contains these species. Plants with seeds dispersed by birds and wind recolonize, but it takes a considerable amount of time. Plants that have undergone regional declines, such as the American yew, hobblebush, native honeysuckle, blue-bead lily, and speckled wood lily, may take many decades to recover on their own. Recolonization or reintroduction is not possible if the deer density remains high.

From 1905 to 1950, the chestnut blight removed the American chestnut as a major source of hard mast for deer and other wildlife throughout much of the Eastern United States, including areas with a high proportion of oaks. At the same time, the forest overstory closed following decades of heavy harvest in many areas and reversion of failed farms to forestlands in others. Over time and across a broad area, food resources for deer in forests declined. Shade-tolerant plants that are not preferred as deer forage, such as hay-scented, New York, and bracken ferns, and a few other species resistant to deer browsing (more recently including nonnative plants such as Japanese stiltgrass and garlic mustard), proliferated and dominated the understory in many forests.

Understanding density and carrying capacity

Deer density fluctuated across the Eastern United States from the late 1800s through the 1960s in response to land use practices. It is important to realize that the deer density responded and differed markedly in various areas of the Eastern United States during this time as a result of different land use practices and habitat quality. However, by the mid- to late twentieth century across the white-tail deer range, a common experience which hunters became accustomed to was an increasingly high deer abundance while remaining largely ignorant of its impact on forest ecosystems. They believed high deer densities were natural and demanded that wildlife agencies adopt management practices, such as reduced or no antlerless deer harvest, to sustain high deer densities.

By the 1990s, scientists determined that deer density could be maintained at different abundance levels (carrying capacities), with differing levels of impact on a forest ecosystem (Figure 17.1). To maintain deer density at a given carrying capacity, mortality factors (deer harvest, predation, starvation, disease, parasites) must offset recruitment (number of fawns surviving in a given year), immigration (deer moving into an area from some other location), and food supply.

The relationships among density, recruitment, forest management, and carrying capacity are not known precisely except in a very few well-researched sites. These relationships differ depending on conditions. Those depicted on the graph in Figure 17.1, should not be applied

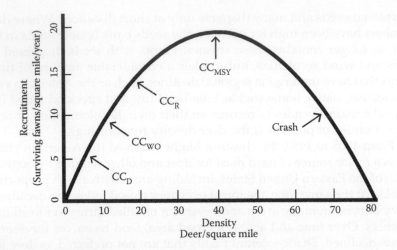

Figure 17.1 Carrying capacities of white-tailed deer in forests. CC_D, carrying capacity for (wildlife and vegetation) diversity; CC_{WO}, carrying capacity for regeneration of oak species in the white oak group; CC_R, carrying capacity for regeneration of most other tree species, including oaks in the red oak group; CC_{MSY}, carrying capacity with the maximum sustainable yield of deer for harvest; crash, density where unsustainable deer density exceeds forage, resulting in massive winter die-off in some regions. Carrying capacities vary geographically and over time; numerical values in a particular area may not match those in this graph, which illustrates the general pattern of deer density in relation to food supply (see text) using research in the Northern Hardwoods Region. The difference between deer carrying capacities for white oaks and for hardwoods in general is large in much of the northeastern part of the white-tailed deer's range and gets smaller or vanishes southward and westward. (Adapted from deCalesta, D.S. and Stout, S.L., *Wildl. Soc. Bull.*, 25, 252, 1997.)

indiscriminately. Deer density interacts with a host of factors that vary from one location to another and from year to year, all working in tandem to affect forest response. The deer density associated with different carrying capacities is affected by such factors as the amount of light that reaches the forest floor, forest type, understory species composition, landscape context, soil conditions, length of growing season, alternative food sources, patterns of seasonal movement by deer, and legacy effects of prolonged high deer numbers. In forests managed for sustained timber production, there is a limit to the proportion of a forested area that can be harvested annually. Generally referred to as the *annual allowable cut*, this upper limit to harvest also represents an upper limit of forage creation for deer, with a corresponding upper limit of carrying capacity for deer.

Several tools and methods are used to estimate deer density, including infrared-triggered cameras, aerial flights with infrared videography, counting and analyzing deer pellet groups, and various computer modeling techniques. However, it is important to realize that there is no measure of accuracy with any method unless a marked segment of the deer herd is being monitored. Thus, it is most important to follow the trends of density estimates over time and monitor the effects of deer density on sites (e.g., quantifying browsing pressure, plant species diversity, etc.).

Carrying capacity for diversity

Deer populations were sustained at a carrying capacity for diversity (CC_D) prior to the European colonization of North America when there was a larger complement of predators and forage production was sporadic and geographically limited, created by infrequent and variably sized disturbance factors, such as tornadoes, tropical storms, ice storms, and fire— either natural or set by Native Americans. Deer density was low (because of constant pressure by the predator community), allowing a vast and diverse assemblage of seedlings, shrubs, and herbs to proliferate. The wildlife community, which increases in diversity with plants and complexity of vegetation structure, is highly diverse at CC_D because of the variety and abundance of ecological niches.

Carrying capacity for tree regeneration

Restrictions on hunting in the early 1900s, coupled with vast amounts of forage created by regrowth following timber harvest and an abundance of agricultural crops, resulted in great increases in carrying capacity, which when combined with restrictions on deer harvest, resulted in exponential increases in deer density. Deer greatly reduced seedlings, shrubs, and herbs in order of their feeding preferences, nearly eliminating many highly preferred species, which may differ by region. Less-preferred species were sufficiently abundant for successful regeneration of a limited diversity of trees and herbs following timber harvest or other disturbances that allowed sunlight to reach the forest floor. This deer carrying capacity is labeled carrying capacity for tree regeneration (CC_R), a carrying capacity that permits regeneration of some tree species. At this level of deer density, there is noticeable reduction in diversity and abundance of plants in the understory, especially herbs and shrubs. Deer prefer various oaks over others. Generally, various species in the white oak group are selected over those in the red oak group. However, once seedlings of preferred species are greatly reduced or eliminated by browsing, deer turn to less-preferred species, including other oak species. Therefore, red oaks should be lumped into CC_R, and depending on regional preferences, white oaks may be associated with a lower deer density, CC_{WO}.

Carrying capacity for maximum sustained yield of deer

Even at and above CC_R, there is still much forage available for deer. In the absence of adequate predation and hunting mortality, carrying capacity reaches the highest sustainable density—carrying capacity for maximum sustained yield of deer (CC_{MSY}). This level of deer abundance, called the *maximum sustained yield* or *nutritional carrying capacity*, is the point where recruitment is highest in terms of absolute numbers. However, continued recruitment at this level is dependent on continued regeneration of forage and/or deer removal. High deer densities at CC_{MSY} and relatively high deer harvests (primarily antlered deer—hunters were ingrained with the philosophy of protecting does to retain high reproduction) framed the hunter experience in various regions from the 1930s to the 1990s. Many hunters came to believe this was a sustainable and desirable situation, representing good deer management. However, continued browsing pressure by abundant deer herds began to reduce forage availability, which was increasingly dominated by plant species less preferred by deer and lower in nutritional value as the best foods were depleted. Vegetation dynamics in the understory changed, and the formerly diverse herbaceous and woody plant community in some regions was compressed into a community dominated by mat-forming ferns, grasses, and tree seedlings resistant to deer browsing. At this point, deer populations exceeded CC_{MSY}.

Quality of deer (body weight and antler characteristics) declines as deer are unable to ingest enough energy for growth or to store adequate fat because of reduced food quality and quantity. Decades of harvesting primarily yearling bucks prevents bucks from reaching older age classes and thus producing relatively large racks. As forage availability increases following timber harvest or other disturbances, CC_{MSY} (and deer quality) may increase. However, higher densities of deer permitted by the additional forage can intensify impacts on tree seedlings and other understory vegetation as well as reduce height and diversity of understory vegetation. Deer and the associated habitat should be managed for CC_{MSY} only when the primary objective is for hunters to harvest a maximum number of deer.

Crash

As forage quality and quantity decline in the face of increasingly and unsustainably high deer abundance, severe winters (in northern latitudes) and increased transmission of parasites and disease can result in massive deer die-offs, or crashes. Because of the poor condition of deer, reproduction and recruitment are low when deer herds crash. It takes several years before deer herds can again build up. In the past, such crashes were followed by hunter demands to close or greatly reduce deer hunting season. When such actions were followed by increases in tree harvest as forests matured to harvest age, deer populations again rapidly increased with increased forage.

The balance between mortality and recruitment of white-tailed deer must be addressed annually by adjusting mortality (primarily by hunting) to prevent deer from exceeding the carrying capacity selected as a goal. If this is not done, deer abundance will exceed the desired carrying capacity, resulting in an unacceptable impact on forest ecosystems, and finally on the deer herd itself.

In many states, hunters remember when deer density was high. They saw many deer during hunting season, and they killed mostly yearling bucks with scrawny antlers. They think such deer abundance is natural, and they are resistant to aggressive harvest regulations favoring an increased doe harvest. On the other hand, foresters and ecologists are aware of the loss of plant diversity, wildlife community diversity, and successful tree regeneration when deer density exceeds carrying capacity for regeneration, especially those species deer consistently prefer. Agencies responsible for managing deer often find themselves between hunters who want deer managed at or above CC_{MSY} and foresters, ecologists, and wildlife management professionals who are aware of the effect of artificially high deer populations and want deer managed at CC_D, CC_{WO}, or CC_R.

Hunters of other species, such as ruffed grouse, often are unaware that for them deer should be managed at CC_D instead of CC_{MSY}. CC_{MSY} can negatively affect habitats to the point where some game and nongame species decline in abundance and others may be unable to persist because cover for them is decimated (e.g., nesting and feeding structures for shrub layer–dependent birds, such as black-throated blue warblers and mourning warblers).

Is there a problem?

To determine whether you need to manage deer to conform to a desired carrying capacity, you must first compare it to the existing carrying capacity. Such an assessment requires an evaluation of the extent and intensity of their impact on understory vegetation. Usually, these measurements are collected from plots placed along transects that are systematically located throughout the forestland (see Pierson and deCalesta 2015 for an example technique). Characteristics of understory vegetation representative of the five carrying-capacity situations (diversity, oak regeneration, some tree regeneration, maximum sustained yield, and deer population crash) are described in Table 17.1.

Table 17.1 Characteristics of forest understory vegetation (forest health indicators) at five deer carrying capacities

Carrying capacity	Vegetation characteristics (forest health indicators)
Diversity (CC_D)	Understory dominance by full complement of native herb, shrub, and seedling species; existence of groundcover (less than 3-feet tall) and shrub cover (3- to 10-feet tall)
Oak regeneration (CC_WO)	Acceptable stocking levels[a] of seedlings and saplings of white oak species; limited presence of mat-forming ferns,[b] nonnative, invasive species, and other plants that interfere with the development of seedlings; herbs and shrubs preferred by deer possibly sparse or absent; open understory with sparse shrub layer
Some tree regeneration (CC_R)	Acceptable stocking levels[a] of seedlings and saplings of desired tree species other than white oaks; limited presence of mat-forming ferns,[b] nonnative, invasive species, and other plants that interfere with the development of seedlings; herbs and shrubs preferred by deer sparse or absent; open understory with sparse shrub layer
Maximum sustained yield (CC_MSY)	Acceptable seedling stocking only of tree species resistant to deer browsing; extremely sparse or absent shrub layer; lack of preferred shrubs and herbs; groundcover either very sparse or consisting mainly of mat-forming ferns[b] or nonnative, invasive species; development of a browse line[c]
Population crash	Well-defined browse line[c]; no vegetation in ground and shrub layers except mat-forming ferns,[b] nonnative, invasive species, and other plants seldom eaten by deer

[a] Number of seedlings counted in plots representative of sufficient numbers to regenerate desired tree species following tree harvest.
[b] Mainly hay-scented fern, but sometimes also New York or bracken ferns, which carpet the ground with single fronds rather than growing in tufts or rosettes like most ferns.
[c] Heavy browsing by deer eliminates vegetation from the ground up to 6 feet, the effective browsing height of deer, discerned as a distinctive line of vegetation with a clearly demarked edge below which there is little or no vegetation.

Is it deer?

In the deciduous forests of the Eastern United States, if there is excessive browsing damage to seedlings, it is almost always caused by deer. Mice, voles, rabbits, hares, and turkeys feed on tree seedlings and soft and hard mast (including acorns), but their numbers are most controlled by predators, including snakes, hawks, owls, shrews, weasels, foxes, coyotes, and bobcats. However, in places where there are sufficient numbers of grazing or browsing mammals, such as cattle, horses, goats, swine, and sheep, seedling regeneration or acorn survival to germination may be reduced significantly and levels of impact characteristic of CC_{WO} and higher may be reached.

Browsing by deer is characterized by a shredded or torn appearance of twigs. Deer have no upper incisors, so they browse off twigs by stripping them between their lower incisor teeth and a tough pad on the upper jaw. Browsing by sheep and goats is similar to that of deer, but they are rarely found in interior forests. In small, isolated woodlots surrounded by pastures, intensity of goat and sheep browsing may be sufficient to reduce or eliminate woody regeneration. Damage by swine (feral hogs) is restricted to rooting for and eliminating acorns, and uprooting of regeneration. Cattle and horses severely impact grasses and herbaceous plants, and also shrubs and regenerating trees. Browsing by rabbits and hares is characterized by a sharp, 45° cut where twigs are clipped off. Browsing by mice and voles is characterized by the shredding of bark low on the seedling near the ground.

On the other hand, if there is scarcity or a total lack of seedlings, it may be caused by the lack of stimulus for germination and growth. The only way to stimulate germination and seedling growth is to open the forest canopy, allowing additional sunlight to reach the forest floor. The disturbance may occur naturally or by humans.

The level of browsing on seedlings should be evaluated to verify whether browsing is causing a lack of adequate seedling regeneration (Figure 17.2). If seedlings of all species are not browsed or are lightly to moderately browsed, enough will survive to produce mature trees. This condition represents CC_D. If seedlings are heavily to severely browsed, they may not survive to produce mature trees. CC_R and CC_{WO} are represented by seedlings of desired species at low to moderate browsing levels. CC_{MSY} is represented by light to moderate browsing of seedlings of resistant species, but the level of browsing on the desired seedling species will be heavy to severe. If seedlings of all species are browsed heavily or severely, the deer herd may crash if steps are not taken to reduce deer density or to exclude deer.

Simplified plant species composition is another more subtle effect of prolonged deer density higher than CC_D. Mat-forming ferns (hay-scented,

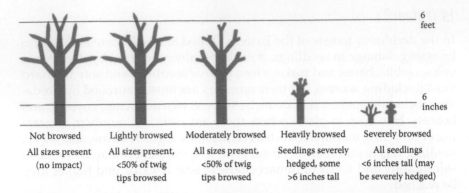

Not browsed	Lightly browsed	Moderately browsed	Heavily browsed	Severely browsed
All sizes present (no impact)	All sizes present, <50% of twig tips browsed	All sizes present, <50% of twig tips browsed	Seedlings severely hedged, some >6 inches tall	All seedlings <6 inches tall (may be severely hedged)

Figure 17.2 Appearance of seedlings at different intensity levels of browsing.

New York, or bracken) or nonnative species, such as stiltgrass or garlic mustard, often dominate the ground layer, and native forbs may be missing entirely. The shrub layer, if it has not been eliminated, is reduced to the least palatable species, such as the striped maple in the northern part of the white-tail deer range, spicebush in the center, pawpaw toward the South, and nonnative shrubs—Japanese barberry, ornamental-escaped honeysuckle, and many others—throughout the range.

Managing deer impact

There are two management approaches for reducing and stabilizing deer impact on forest ecosystems: (1) reduce deer density or (2) protect understory vegetation from excessive deer browsing using exclusion devices or repellents.

In situations where deer density has been high for a long time—15 years to as many as 75 years in some areas—long-lasting, chronic effects of overbrowsing can cause prolonged delays in forest recovery unless the vegetation is managed. Legacy effects of prolonged high deer numbers include inadequate seed availability and dominance by vegetation that interferes with regeneration of native trees, shrubs, and herbaceous understory plants.

Reducing deer density

Hunter harvest: Regulated hunting is the most effective strategy to reduce deer density to the desired levels. Harvesting adequate numbers of antlerless deer is a proven way to increase overall deer harvest and reduce population density. Because many hunters primarily hunt for antlered deer, they must be motivated to harvest additional antlerless deer, which is essential for reducing deer density and maintaining it at desired levels. Deer management by recreational hunting requires managing hunters.

For optimum effectiveness in achieving CC_D, CC_{WO}, or CC_R, managing deer impact with regulated hunting requires several steps:

- Characterize the status of appropriate indicators of forest health for the chosen deer carrying capacity (see Table 17.1).
- Determine whether deer abundance is just right, too high, or too low, depending on the condition of forest health indicators (presence/absence of various plant species) for the chosen deer carrying capacity.
- Experiment with adjusting hunting pressure on antlerless deer if the condition of indicators and deer abundance does not match the desired level of forest health and deer carrying capacity.
- Provide and maintain good access for hunters.
- Establish a trusting and effective relationship with hunters, including two-way communication that provides hunters with information they want and need (deer density and health) and signals that their input is genuinely considered in deer management decisions.
- Deer management strategies must be transparent and supported by science.
- Promote the benefits to hunters when deer density is reduced (healthier deer, better opportunities to hunt other game species, active participation in science-based management to sustain good hunting over the long term).

Deer culling by sharpshooters: Sometimes it is not possible, for safety or other reasons, to allow public hunting to reduce deer density. It is not uncommon that increased deer harvest by hunters proves ineffective to reach management goals for tree regeneration or recovery of other plant species or forest ecosystem functions. Professional marksmen can remove deer, usually at night over bait with rifles fitted with silencers, with minimal disturbance or risk to nearby homeowners. The carcasses are removed, and the meat is usually distributed to charitable food banks. The cost of hiring marksmen is high and a real consideration for townships or other agencies trying to reduce deer density. Permits must be obtained from the state wildlife management agency to allow sharpshooting. Not all agencies will issue such permits, and those that do will issue permits only if a convincing case can be made that the local conditions prevent regulated hunting from lowering deer density sufficiently to meet ecological goals. For these reasons, sharpshooting is generally restricted to urban and suburban parks and nature preserves. Even so, local hunters often complain this activity diminishes hunting opportunities. Sharpshooting can reduce deer density much faster than hunting. Some managers contract for one or a few years of sharpshooting in hopes that after the deer population has been drastically reduced, switching to managed hunting will be sufficient to keep the population at the lower level.

Predation by "natural" predators: A suite of large predators, including bears, wolves, mountain lions, coyotes, bobcats, and Native Americans, maintained a much lower deer density than those we see across much of the species' range today. Black bears, coyotes, and bobcats are the only "natural" predators remaining in the Eastern United States. The effect of predation is not usually sufficient to offset recruitment in many areas because of a lack of predators. However, in some areas of the Eastern United States, predators have an impact on deer density and recruitment.

Trapping and transplanting: Theoretically, deer may be trapped and transplanted to other places, but the "other places" then may experience problems caused by too many deer. In most states, trapping and transplanting have been phased out in recent years. Willing recipients are limited, the procedure is costly, and it must be repeated continually to be effective. Because the practice also increases the risk of disease transmission, state wildlife regulations typically ban trapping and transplanting except in rare instances in which the capture of small numbers may be approved for research.

Chemosterilants: Chemosterilants are an attractive concept for those who do not want deer populations reduced by hunting but prefer the idea of birth control. Unfortunately, technology is not at the stage where its use is effective or feasible in most deer populations. Many state wildlife agencies prohibit the use of chemosterilants on wild populations except on a small scale in approved research. In methods currently undergoing research and testing, female deer are injected with a chemical that prevents reproduction. Some of the chemicals may be placed in syringes fired from a dart gun, with an effective delivery range of about 100 feet. Other chemosterilants must be injected by hand, requiring deer be captured and restrained prior to injection. Average cost ranges from $100 to $500 per deer per injection. For some of the chemosterilants, does must receive additional "booster" injections to remain infertile in subsequent years. A majority of does in a herd must be injected to effect meaningful reduction in reproduction and recruitment.

Chemosterilants do not result in instant reduction in deer herds; rather, they result in a slowing of reproduction and recruitment. The logistics of injecting enough does to reduce reproduction and recruitment significantly render chemosterilants potentially useful only on small (less than 500 acres) islands or landscapes that are fenced to preclude immigration of unsterilized does. For a chemosterilant to be an effective method for reducing deer density, it must be developed that only deer (and no other wildlife species) are rendered infertile, and that it is effective when ingested, so that it can be administered over large areas via bait stations. The United States Department of Agriculture Wildlife Services, the federal agency charged with research on methods of reducing animal-caused damages, states chemosterilants are not a replacement for lethal methods in managing deer populations.

Protecting understory vegetation

Increasing deer forage: Thinning or other timber harvest methods can be used to increase deer forage availability. The increase in forage typically persists for 5–7 years before tree seedlings grow beyond the reach of deer. Forest management should be spread out, both spatially and temporally, to maintain forage availability across time and space. It may be necessary to treat 10–40 acres to temporarily overwhelm deer and allow more susceptible seedlings and herbaceous plants to establish. Deer may respond to increasing levels of forage following timber harvest with higher recruitment rates, necessitating increased mortality rates from hunting to prevent an increase in deer density. If the increased forage production cannot be maintained through sustainable forest management or deer density is allowed to increase, the result will be additional severe impacts on forest understory, other wildlife and their habitat, and the health of the deer herd itself. Thus, any forage increases must be sustained and coupled with an appropriate deer harvest to prevent deer density and impacts from increasing to unacceptable levels.

Other methods of increasing deer forage, such as developing and maintaining food plots planted to forage species such as clover, brassica, and chicory, are used by managers to offset deer-foraging pressure on the forest understory, while efforts are under way to reduce deer density and improve plant diversity. Various habitat management practices are promoted by the quality deer management philosophy that focuses on balancing deer density with improved habitat. In addition to harvesting an appropriate number of antlerless deer annually and restricting buck harvest to older deer, the approach emphasizes managing habitat (judicious use of timber harvest, prescribed fire, selective herbicides, and supplemental plantings) to improve available nutrition and increase deer quality (increased body weights, improved antler characteristics). The Quality Deer Management Association recommends maintaining deer density at or below CC_{MSY}.

Fencing: Fencing may exclude deer from forested areas recently regenerated. Ten-foot high woven wire or livestock fencing works much better than electric fencing, which requires more maintenance. Also, deer can easily penetrate electric fences and are difficult to remove from fenced areas. Woven wire fences, which are designed to last for 5–10 years generally cost at least $2 per linear foot to build, plus annual maintenance costs. More durable fencing, typically designed to last 25 years or more, may also exclude deer, but initial construction and annual maintenance costs are very high. Falling limbs and trees can knock down woven wire fencing, so trees along both sides of the fence should be cut back a sufficient distance. The cleared zones can double as firebreaks where the forest is managed with prescribed fire, and to allow passage of all-terrain vehicles for fence monitoring and repair. Deer impact on understory vegetation within fenced areas must be monitored to detect whether deer have penetrated the fence and have begun to impact plant diversity.

Tubing: Oak seedlings can be grown in nurseries and planted inside protective, degradable plastic tubes (generally 4 feet tall and staked to the ground), which prevent deer from browsing on the seedlings until they develop enough height and diameter growth to escape deer browsing and also withstand wind and antler rubbing. Tubes and nursery-grown seedling stock are relatively expensive. Also, saplings grown inside tubes often are spindly and may be susceptible to wind damage, resulting in regeneration failures after the tubes decompose or are removed. Additionally, the only resources protected are oak seedlings planted inside the tubes, as deer are free to eat all other understory vegetation.

Repellents: In areas where regeneration is achieved by planting nursery stock (most often of coniferous trees), both taste and area repellents have provided some protection. Taste repellents are applied to the tree foliage. However, precipitation washes off repellents, which then must be reapplied. Additionally, the repellent value diminishes over time as deer become accustomed to the taste or odor.

Counteracting effects of prolonged high deer density on understory vegetation

Inadequate seed availability: After reducing the deer population and establishing a program to maintain it at that level, intervention may be needed to hasten the return of missing understory shrub and herb species. Planting nursery stock is the quickest way to reestablish plant populations of desired species whose seeds are unlikely to be present or returned by natural means.

Interfering vegetation: Dense cover of mat-forming ferns or other undesirable plants that proliferate where deer have eliminated their competitors can be an impenetrable barrier against establishment of new trees, shrubs, or herbaceous plants. Dense shade and root competition suppress germination and overwhelm the few seedlings that manage to emerge. Application of herbicides (see Chapter 10) combined with partial overstory removal, such as shelterwood harvest or thinning (see Chapters 7 and 11), allows for germination and growth of tree seedlings if deer density is at or below CC_R.

Selecting a management strategy

Managing deer impact on oak regeneration requires consideration of management goals, forestland size, and the composition of the surrounding landscape, along with appropriate management tools. Table 17.2 is a guide for the selection of appropriate management strategies.

There are three alternative forest management goals that relate to deer carrying capacity: (1) managing to optimize diversity and abundance of all forest resources, including deer health (CC_D); (2) managing to sustain timber harvest (CC_R or CC_{WO}) for a suite of commercially desirable tree species and coincidentally improve habitat for a limited suite of plant and wildlife species, including white-tailed deer; and (3) managing to optimize deer harvest quantity or quality (CC_{MSY}).

Table 17.2 Guide for selecting a management strategy to achieve desired carrying capacity and deer impact level

Carrying capacity	Landscape context	Management strategy
CC_D, CC_{WO}, CC_R	Urban/suburban[a]	Fencing plants and sensitive areas; sharpshooting; strictly regulated hunting
CC_{MSY}	Urban/suburban	None—too much public conflict and hazard
CC_D, CC_{WO}, CC_R	Agriculture and forest (small woodlots[b])	Regulated hunting; forest management; food plots; fencing plants and small openings
CC_{MSY}	Agriculture and forest (small woodlots)	Regulated hunting (however, CC_{MSY} generally not a goal because of conflicts with agriculture); forest management; food plots
CC_D, CC_{WO}, CC_R	Forestlands	Regulated hunting; forest management; food plots; fencing regeneration sites
CC_{MSY}	Forestlands	Regulated hunting; forest management; food plots

[a] For hunting within urban/suburban landscapes, including parks, hunters must be intensely supervised to ensure no adverse hunter–general public interaction, such as trespassing, shots fired across residences, dragging deer carcasses across residential properties, etc.

[b] Most agriculture and small forest areas are owned by persons practicing small-scale farming with isolated, small woodlots wherein little timber management occurs and trees are harvested occasionally to produce cash for emergencies.

There are three landscape categories wherein deer and forests are managed: (1) contiguous blocks of forestland comprising hundreds to thousands of acres; (2) isolated blocks of forestland interspersed within a larger agricultural landscape; and (3) isolated blocks of forestland surrounded by urban/suburban landscapes.

Monitoring: Have you solved the problem?

Monitoring the impact of deer foraging on the forest understory over time is essential to determine if the management practices used were successful. If the program employed was not successful, examine your strategy, determine the problem, and make appropriate adjustments.

Effective management is adaptive. Learn from experience and modify behavior as a result of that experience. Think of it as a series of feedback loops where current knowledge is applied to solve a problem, and the results are monitored to rate success. Adaptive management employs incremental steps (monitor, adjust, monitor) until success is achieved and maintained.

Further suggested reading

Beaver, J.T., C.A. Harper, R.E. Kissell, Jr., L.I. Muller, P.S. Basinger, M.J. Goode, F.T. van Manen, W. Winton, and M.L. Kennedy. 2014. Aerial vertical-looking infrared imagery to evaluate bias of distance sampling techniques for white-tailed deer. *Wildlife Society Bulletin* 38(2):419–427.

Curtis, P.D., B. Boldgiv, P.M. Mattison, and J.R. Boulanteger. 2009. Estimating deer abundance in suburban areas with infrared-triggered cameras. *Human-Wildlife Interactions* 3:116–128.

deCalesta, D.S. 1997. Deer and ecosystem management. In W.J. McShea, H.B. Underwood, and J.H. Rappole (eds.), *The Science of Overabundance: Deer Ecology and Population Management.* Smithsonian Institution Press, Washington, DC, pp. 267–279.

deCalesta, D.S. 2013. Reliability and precision of pellet-group counts for estimating landscape-level deer density. *Human-Wildlife Interactions* 7:60–68.

deCalesta, D.S. and S.L. Stout. 1997. Relative deer density and sustainability: A conceptual framework for integrating deer management with ecosystem management. *Wildlife Society Bulletin* 25:252–258.

Goode, M.J., J.T. Beaver, L.I. Muller, J.D. Clark, F.T. van Manen, C.A. Harper, and P.S. Basinger. 2014. Capture-recapture of white-tailed deer using DNA from fecal pellet groups. *Wildlife Biology* 20(5):270–278.

Halls, L.K. (ed.). 1984. *White-tailed Deer: Ecology and Management.* Stackpole Books, Harrisburg, PA, 870 pp.

Hewitt, D.G. 2011. *Biology and Management of White-tailed Deer.* CRC Press, Taylor & Francis Group, Boca Raton, FL.

Lashley, M.A., C.A. Harper, G.E. Bates, and P.D. Keyser. 2011. Forage availability for white-tailed deer following silvicultural treatments. *Journal of Wildlife Management* 75:1467–1476.

Latham, R.E. (ed.), J. Beyea, M. Benner, C.A. Dunn, M.A. Fajvan, R.R. Freed, M. Grund, S.B. Horsley, A.F. Rhoads, and B.P. Shissler. 2005. Managing white-tailed deer in forest habitat from an ecosystem perspective: Pennsylvania case study. Audubon Pennsylvania and Pennsylvania Habitat Alliance, Harrisburg, PA, 340 pp.

McShea, W.J., H.B. Underwood, and J.H. Rappole (eds.). 1997. *The Science of Overabundance, Deer Ecology and Population Management.* Smithsonian Books, Washington, DC, 394 pp.

Miller, K.V. and R.L. Marchinton (eds.). 1995. *Quality Whitetails: The Why and How of Quality Deer Management.* Stackpole Books, Pittsburgh, PA, 320 pp.

Minnesota Department of Natural Resources. (No date.) *Fencing Handbook for 10' Woven Wire Deer Exclusion Fence.* Minnesota Department of Natural Resources, Wildlife Damage Management Program, Brainerd, MN, 22 pp.

Pierson, T.G., and D.S. deCalesta. 2015. Methodology for estimating deer browsing impact. *Human-Wildlife Interactions* 9:67–77.

Potvin, F. and L. Breton. 2005. Testing two aerial survey techniques on deer in fenced enclosures: Visual double-counts and thermal infrared sensing. *Wildlife Society Bulletin* 33:317–325.

Quality Deer Management Association. 2010. Living with white-tailed deer: A community guide to urban deer management. Quality Deer Management Association, Bogart, GA, 54 pp.

Stolzenburg, W. 2008. *Where the Wild Things Were: Life, Death and Ecological Wreckage in a Land of Vanishing Predators.* Bloomsbury, New York, 291 pp.

chapter eighteen

How do I get started?

Larry A. Tankersley

Contents

The basic plan 279
Research your property 280
The state of the forest 281
Combine and analyze the information 282
Getting help 282

The most efficient way to handle all the information contained in this book is with a written management plan. The plan is a guide to help you make decisions based on facts. Without a plan, management often lacks direction and may be haphazard at best. A management plan is your blueprint for developing the forest you want.

Forest management plans are common for industry and government land. Many owners of family forests also recognize the value of their written management plans in assuring that the right things get done in a timely manner. Forest management plans can be simple with very little detail or they can be complex and detailed. Whether the plan is simple or complex, it should be in written form.

The basic plan

A basic forest management plan consists of (1) a statement of goals, objectives, and expectations from forest management; (2) a land description; (3) a state of the forest report; (4) recommendations for proposed management and a timeline for proposed activities; and (5) an appendix.

Written goals and objectives are necessary for effective management planning. The statement of ownership goals and objectives and your forest management philosophy should reflect your wishes. Increased income may or may not be your objective for every acre. Forest management decisions are usually the result of choices made among the benefits of several alternative uses (see Chapter 13). A final statement of goals and objectives may not be possible until the management plan is near completion and you have had time to gather more information on your forest's potential.

A management plan should contain a detailed description of the property. This section of the plan should include maps and/or aerial photos and written descriptions of the specific location and land boundaries of your property.

An ownership map, made from the legal description of the property, should show the directions and length of each boundary line and should indicate how that line is currently marked (fence, old paint, flagging, etc.). The map might also show proximity to nearby towns, public roads, and other noteworthy locations, such as cemeteries or churches.

Research your property

In preparing this information, it may be necessary to work at the county or parish courthouse. It is extremely important to know your rights to minerals, water, and other legal aspects of landownership, which are particular to your county and state. These should be in your written plan. A deed abstract, which lists all the previous owners of your land, is a first step in developing a history of your property.

Copies of property deeds, lists of adjacent landowners, and a boundary maintenance schedule are useful additions to your ownership information.

In describing you land, it can also be helpful to include information about climate, rainfall, and length of the growing season. Special weather information such as the probability of ice, high winds, floods, and droughts may be useful in making decisions.

Topographic information is necessary for planning management and harvesting operations. Topographic maps are helpful in planning access, firebreaks, managing smoke, and other general purposes. These are available from the U.S. Geological Survey, and digital versions of these maps can be downloaded for free from their website (http://topomaps.usgs.gov/drg/).

No land description is complete without a soils map. A soils map is helpful in delineating areas of equal site quality and judging land use capabilities and limitations. Soil maps for most areas are available from the Natural Resources Conservation Service (NRCS) and in many cases can be accessed online at the NCRS Web Soil Survey website (websoil survey.nrcs.usda.gov).

Aerial photos or high-resolution satellite imagery of your property are invaluable to the development of a good management plan. Photos or satellite imagery should be at a high resolution (e.g., 1 m resolution), which means they will easily show details, such as individual trees, buildings, and roads in high-resolution images. Imagery at a coarser resolution (e.g., 30 m resolution) will have a grainy or blurred appearance and it will be difficult to distinguish individual trees or buildings. Aerial photos are often referred to as *digital orthophoto quarter quads* (DOQQs). If you have a computer, free computer mapping programs are available that you may use to make your own maps. One of the most popular is Google Earth™ (www.google.com/earth/index.html). It uses high-resolution satellite imagery and also includes symbols to highlight other features, such as roads and streams. These programs allow you to zoom into any location on the landscape and also allow you to save and print the image on the screen, so you can zoom into your property and create your own property map. They can also communicate with recreational GPS units. If you own a GPS unit, you can use it to collect information around your property, such as boundary points or locations of unique features, and then download those onto the map. Much of the information necessary for a workable plan can be obtained quickly and accurately from an aerial photo. Old photos can be used to help establish a history of your land and recent photos provide an overall picture of your land in its current condition. Aerial photos can be obtained from the local Farm Services Agency (FSA), NRCS office, state forester, or from a private aerial photo service.

The state of the forest

The state of the forest section of your plan should describe in detail the forest occupying your land. This section should contain a forest type map, which delineates areas where similar species associations occur. Forest types may include upland pine, oak/pine, oak/hickory, bottomland hardwood, and cove hardwoods (in mountainous areas), among others (see one example of forest types in Chapter 14).

The type map should show existing and proposed roads, streams, gullies, fences, easements, dwellings, and impoundments. Special features, such as rock outcrops, steep slopes, or old home sites that you might want to preserve, should also be included on the map.

In addition to your forest type map, you will need a stand map. A stand is a basic mapping unit defined as a contiguous area of one timber type with similar origin, history, site quality, and operability. Once delineated, forest stands typically become management units in your plan—each to be treated individually. Tables summarizing the acreage in each stand are useful in evaluating your management options.

Another major component of your state of the forest report is an inventory of present conditions in each forest stand. The species, size and age distributions, stocking, site quality, growth rates, and merchantable volume of the standing trees should be part of the forest inventory. Growth projections are often included in this section. An inventory of your oak regeneration pool (seedlings and saplings and their competitive position—overtopped, free to grow) is helpful as well. Tables summarizing the inventory are the best means of presenting this information.

Other site conditions, such as indications of insects or disease, indications of animal use or deer damage, the types and amounts of vegetation wildlife species can use for food and/or cover, and descriptions of development potential may be useful in the inventory. Any information measuring your forest's potential for any of a variety of uses should be included in your state of the forest report.

Combine and analyze the information

The next section of your plan combines all the information you have assembled thus far into a plan for the future. This part of the plan consists of discussions regarding proposed treatments for each stand in your forest. Recommendations should follow from your state of the forest report and the suggested activities should move you toward your desired future conditions. Recommendations should be flexible and allow for changes in your management objectives and timber markets. This discussion should also provide a timeline and budget to ensure management is conducted in a timely fashion.

Prescriptions for specific tasks, such as prescribed burns, herbicide applications, thinning and preparations for regeneration, should be included in this section of your management plan. Plans for forest protection, harvesting, sales policies, and economic considerations should also be covered.

The final part of your management plan is an appendix. This section is for things repeatedly referred to throughout the rest of the plan. Definitions of terms and commonly used information concerning things such as thinning methods, descriptions of insects and diseases, sample sales contracts, and volume tables should be placed in the appendix. Lists of adjacent landowners, vendors, consultants, state foresters, timber buyers, and other sources of information and services can be quickly available if they are in the appendix of your written management plan.

A written management plan is an invaluable tool for getting started in achieving your forest management goals. Without a plan, forest management is often inconsistent and unsatisfactory. A workable plan will help you decide what to do, as well as when and how to do it. Always seek professional assistance when preparing a forest management plan.

Getting help

Get assistance from people focused on your interests to help you develop long-term goals and objectives. Your local Extension Office should refer you to sources of free unbiased information and advice. Also, see Chapter 12 for a comprehensive list of professional resources that may be available to assist you in developing your forest management plan.

Index

A

Accord®, 140–141
Acorns and oaks
  black bears, 42–43
  eastern wild turkeys,
    44–45
  nongame forest birds, 45–46
  production characteristics, 42
  ruffed grouse, 45
  small mammals, 44
  unsuitable substitute,
    46–47
  white-tailed deer, 43–44
Aerial sprayers, 145–146
Allegheny Hardwoods, 27, 29
All-terrain vehicle (ATV), 122, 145, 196
American Forestry Association, 103
Appalachian Hardwoods, 29–30
Arsenal AC®, 140–141
Artificial regeneration
  acorns and nursery seedlings, 86
  afforestation efforts, 86
  bottomland oak seedlings, 89
  direct seeding (see Direct seeding)
  fertilization and weed control,
    96–97
  investments vs. goals, 97
  nursery seedlings (see Nursery
    seedlings)
  planting program, 88
  seedling protection, 95–96
  seed-producing mature trees, 85
  seed sources, 89–90
  site and species, 86–87
  supplementing reasons,
    85–86
Atmospheric stability, 111, 113, 126
ATV, see All-terrain vehicle

B

Backing fire, 124–125
Backpack mist blowers, 145
Best management practices (BMPs), 196,
    212, 251
Big Piney Ecosystem Restoration
    Project, 242
Black bears
  acorns and oaks, 42–43
  denning, 47
  fire, 115–116
  shelterwood method, 80
BMPs, see Best management practices
Bottomland
  acorn production, 258
  animal competition, 252–253
  bole quality, 257–258
  cherrybark oak, 33–34
  coastal plain, 36–37
  ecology, 248
  flooding, 33
  floodplain systems, 36
  forest structure and diversity, 252
  landscape, 251–252
  management considerations, 253
  Nuttall oak, 33, 35
  Shumard's oak, 33–34
  species/site relationships
    alluvial deposits, 250
    disturbance regimes, 251
    floodplain (minor and major), 249
    ridge and swale topography, 248
    riverfront hardwood species, 250
    sediment deposition, 249
    topographic variations, 249
    willow and Nuttall oaks, 250
  stand development, 250
  topography, 33, 36

C

Carrying capacity
  crash, 266–267
  density, 263
  forest understory vegetation, 267–268
  maximum sustained yield, deer, 266
  populations, 265
  tools and methods, 264–265
  tree regeneration, 265
  vegetation dynamics, 266
Catoosa Savannah restoration project,
      243–244
Central Hardwood Region
  Allegheny Hardwoods (*see* Allegheny
      Hardwoods)
  Appalachian Hardwoods (*see*
      Appalachian Hardwoods)
  oak–hickory forests, 25
  oak species, 26
  Ohio Valley, 27
  Ozark/xeric upland hardwoods (*see*
      Ozark/xeric upland hardwoods)
Charcoal iron industry, 11
Clarke–McNary Act, 103
Clearcut method, 77–78
Coastal Plain oak–pine
  loblolly pine/hardwood, 38–39
  oak/pine forest types, 38–39
  silviculture, 36
  Southern United States, 36
  upland oaks, 38
Competition control
  herbicide treatments (*see* Herbicide
      treatments)
  mechanical treatments
    advantages, 134
    disking, 134
    mowing and cultivation, 135
    root raking, 134
  plantation establishment, 133
  prescribed burning, 135–136
  principles, 132–133
  seedlings, 133
  site preparation, 133
Conservation Reserve Program (CRP), 247
Cooperative Extension Service, 179–180
Crop tree release (CTR), 169, 214, 257
  competing trees, 156
  dead trees, 156
  diameter growth, 155
  species diversity, 155–156
  treatments, 156

CRP, *see* Conservation Reserve Program
CTR, *see* Crop tree release

D

DBH, *see* Diameter at breast height
Deer impacts
  carrying capacity (*see* Carrying
      capacity)
  density reduction
    chemosterilants, 272–273
    culling, sharpshooters, 271–272
    hunter harvest, 270–271
    predators, 272
    trapping and transplanting, 272
  in eastern oak forests, 262
  germination and growth, 269
  management strategies, 275–276
  monitoring, 276
  recolonization/reintroduction, 263
  seed availability, 274
  seedling regeneration, 269–270
  understory vegetation protection
    fencing, 273–274
    forage, 273
    repellents, 274
    tubing, 274
  vegetation, 274–275
  white-tailed deer, 261
Diameter at breast height (DBH), 82, 171,
      184, 203, 233, 235
Digital orthophoto quarter quads
      (DOQQs), 281
Direct seeding
  acorns, 92–93
  acorn weevils, 91
  advantages, 90
  disadvantages, 91
  germination, 91–92
  mammals and birds, 91
DOQQs, *see* Digital orthophoto quarter
      quads
Dormant-season fire, 110–111, 115,
      118–119, 242

E

Eastern oak forests
  building material and
      firewood, 11
  chestnut, 13
  commercial forestry, 11
  fire, oak savannahs, 9–10

forest fires, 14
hardwood, 7
logging and land clearing, 15
mesophication, 15–16
and pine, 10
pollen and charcoal, 9
presettlement oak forest types, 7–9
tree species, 14
white-tailed deer, 14
wildfires, 12
Eastern wild turkeys, 44–45
Ecosystem restoration, 188

F

Fertilization and weed control, 96–97
FIA project, *see* Forest Inventory and
    Analysis project
Fire
    air temperature, 112
    atmospheric stability, 111, 113, 126
    black bear, 115–116
    burn plan, 122
    Clarke–McNary Act, 103
    controlled burning, 103
    dormant-season fire, 110
    in eastern oak systems, 104–105
    fencing, 101
    forestry profession, 103
    forest soils and water sources, 107–108
    forest songbirds, 118–119
    frequency, 110–111
    growing-season fire, 110–111
    heating period, 109
    human perception, 128–129
    humidity, 112
    intensity, 109–110
    livestock, 120–121
    mammal community, 116
    management efforts, 113
    northern bobwhite, 117–118
    occasional fire, 114
    old-field and shrubland songbirds, 119
    post-burn evaluation, 127–128
    prescribed fire, 105, 121
    rainfall, 113
    regeneration, 114
    reptiles and amphibians, 119–120
    ruffed grouse, 117
    site preparation, 122
    species and oaks, 101–102
    steady winds, 112
    upland hardwoods, 104

upland oaks, 101
vegetation and wildlife habitat
    canopy, 106–107
    composition and structure, 105
    food availability, 106–107
    soft mast production, 106
    sunlight, 107
weather patterns and forecasts, 111–112
white-tailed deer, 115
wildfire, 105
on wildlife, 108–109
wild turkeys, 116–117
Fire bird, 117–118
Firing techniques
    backing fire, 124
    flanking fires, 125
    point-source fires, 125
    ringfires/center fires, 125–126
    strip-heading fire, 124
Fixed-wing aircraft, 145, 147
Flanking fires, 125
Forest inventory and analysis (FIA) project,
    3–4
Forest management plans
    additional information, 195
    assistance, 282
    attributes, 195
    endangered species, 196
    herbicide applications, 282
    ownership, 279–280
    prescribed burns, 282
    property, 280
    soil surveys, 195
    topographic maps, 195
    treatments, 282
    type map, 281–282
Forest songbirds, 46–47, 80, 118–119
Fuel loads, 104–105, 229, 235, 240, 243

G

Garlon 3A®, 141
Google Earth™, 281

H

Healthy oak–pine communities,
    Arkansas, 243
Herbaceous ground layer, 224–225
Herbaceous weed control
    chemical site preparation, 148
    Goal 2XL®, 149–150
    mechanical methods, 147

oak planting, 147
Oust XP® controls, 149
Select®/Fusilade DX®, 150–151
Herbicide treatments
    aerial sprayers, 145–146
    ATV and RTVs, 145
    backpack mist blowers, 145
    basal sprays and wipes, 143–144
    brand name and common name, 136
    broadcast applications, 144
    cut-treat, 141
    efficacies, species, 141–143
    environment, 136
    foliar sprays, 143
    label, 138
    manufacturers, 136–137
    site preparation, old-field sites, 146–147
    skid and trailer-mounted sprayers,
        144–145
    stem injection (see Stem injection)
    timing, 138–139
    tractor-mounted sprayers, 145
    and wildlife, 137–138
    woody plant control, 139

I

Intermediate treatments
    cleanings, 154
    CTR (see Crop tree release)
    high-grading, 162–163
    improvement cutting, 154
    liberation cutting, 154
    midstory removal, 157
    precommercial thinning, 154
    salvage cutting, 154
    sanitation cutting, 154
    stand improvement, 154
    thinning, 158
    weeding, 153–154

L

Land management, 128, 189–190
Lower Mississippi Alluvial Valley
        (LMAV), 32

M

Mature forests, 255–256
Mature mixed-oak forests
    American chestnut, 204
    dead trees, 204

mast-producing trees, 207
old-growth stand, 207–208
ownership, 204
patchwork, 208
predation, 205
present stand, management,
        212–213
sawtimber, 207
short-and long-lived oaks, 207
site quality, 205
species composition, 205
stand density/stocking level, 206
sub-canopy layers, 205
virgin and old-growth forests,
        203–204
Middle-aged mixed-oak forests
    assessment, 214–215
    crop tree, 215–216
    description, 214
    fire, 217
    insect and disease control, 217
    invasive plant control, 217
    management, 214
    vine control, 216

N

National Woodland Owner Survey, 189
Natural Resources Conservation Service
        (NRCS), 180, 280
Nongame forest birds, 45–46
Northern hardwood region
    disturbance, 24–25
    Lakes States and northeast, 24
    leaf coloration, 23
    mesophytic, 24
    oak-growing regions, 22–23
    oak species, 24
Northern red oak-mixed hardwood type
        group
    combinations, 202
    high-value products, 203
    lower-slope positions, 203
NRCS, see Natural resources conservation
        service
Nursery seedlings
    advantages, 93
    bare roots, 94
    root systems, 93–94
    shovels and sharpshooters, 95
    size and quality, 93
    spacing and stocking, 95
Nuttall oak, 33, 35, 56, 88, 250

O

Oak forests
    acorn production characteristics, 42
    competition control (*see* Competition control)
    geographic distribution, 22
    intermediate treatments (*see* Intermediate treatments)
    landowners, 48
    mature stands, 47–48
    proper and active management, 47
    red and white oaks, 41–42
    regeneration (*see* Regeneration)
    silviculture (*see* Silviculture)
    species, wildlife, 47
Oak–hickory forests, 25, 30
Oaks (upland)
    flexibility, 193
    high-value forest products, 194
    ice/snow/wind damage events, 193
    professional experience, 193
Oak Savannahs and Prairie Region, 38, 40
Ohio Valley
    red maple, 27, 29
    white oak/red oak/hickory forest type, 27–28
    yellow poplar/white oak/northern red oak, 27–28
Ozark/xeric upland hardwoods
    post oak/blackjack oak forest, 31
    southern Missouri and northern Arkansas, 30–31
    sunlight, reproduction, 32

P

Pathway®, 141
PGS, *see* Preferred growing stock
Phytophthora ramorum, 4
Point-source fires, 125
Preferred growing stock (PGS), 161
Prescribed burning, 135–136

Q

Quality Deer Management Association, 273

R

Recreational-type vehicles (RTVs), 145
Regeneration
    advance regeneration, 69
    chestnut oak–post oak, 209

    dormant-season fire, 242
    economic implications, 208
    even-age and two-age methods, 211
    group selection, 210
    growing-season burns, 242
    historic and current, 66–67
    management, 70–71
    principles, 64
    reproduction, 64–65
    residual trees, 211
    seedlings and saplings, 241
    seedling sprout, 209
    shelterwood, 211
    silvicultural treatments, 63
    site quality, 68
    stump sprouts, 69–70
    sub-canopy species, 209
    survey, 210
    white oak–black oak, 210
Restoration (woodlands and savannahs)
    canopy, 235
    contractor, 237
    fire (burns)
        canopy, 239
        dormant-season fire, 239
        germination, 238
        growing-season burns, 239
        herbaceous groundcover, 238
        limbing/topping trees, 238
        stems and rootstocks, 239
        woody species, 238
    fire-tolerant species, 236
    herbaceous groundcover, 235–236
    herbicide, 235
    low-value trees, 237
    residual stems, 235
    shorter-lived species, 237
    site picking, 232–233
    stocking levels, 233
    two-aged structure, stand, 234
    wolf trees, 237
Ringfires/center fires, 125–126
Roundup®, 140, 151
RTVs, *see* Recreational-type vehicles
Ruffed grouse, 45, 47, 78, 80, 117, 267

S

Scarlet oak, 24, 29, 31, 38, 52–53, 200, 211, 213
Seed bank, 105–106, 148, 232, 243, 263
Seedling protection, 95–96
Seedling regeneration, 269–270

Shelterwood method
    harvesting, 78–79
    herbicide treatment, 81
    regeneration, 79, 81–82
    residual trees, 81
    shade and protection, 79
    on wildlife, 80
Silviculture
    canopy gaps, 253–254
    characteristics, 53
    dead and dying trees, 254
    fire tolerance, 58–59
    and forest management, 53
    growth rate, 59
    hardwood forests types, 51–52
    insects and diseases, 60
    longevity, 59–60
    mid-aged bottomland oak forests,
        255–256
    regeneration, 254–255
    seed production, 55–56
    shade tolerance, 57
    soil drainage and flooding, 58
    sprouting ability, 56–57
    sunlight, 254
    thinning bottomland hardwood
        forests, 256
    tree distribution, 53
    tree species, 53–55
    woody debris, 254
Site quality
    evaluation, 197
    mixed-oak forests, 197
    and oak regeneration, 68
    species composition, 20
    topographic and landform features,
        197–198
    tree species composition, 199
Smoke management considerations
    burning, 126
    burn plan, 122
    moisture, 126
    small test fires, 127
    weather and fuel, 111–112
SMZ, *see* Streamside management zones
Southern Hardwood Region
    bottomland forests (*see* Bottomland)
    climate, 32
    LMAV, 32
    oak–pine sites, 33
    upland oak, 33
State forestry agencies, 179
State wildlife agencies, 179

Stem injection
    Accord® and Roundup®, 140
    Arsenal AC®, 140
    chopper, 140
    Garlon 3A®, 141
    Pathway®, 141
    Tordon 101®, 141
    Tordon 101R®, 141
    Tordon RTU®, 141
Streamside management zones (SMZ), 251
Strip-heading fire, 124
Sudden oak death, 4

T

Tax and estate planning, 189
Thinning
    characteristics, 158
    growing stock, 160
    guidelines, 158–159
    logging, 160
Timber objectives
    logistical and financial considerations,
        185, 187
    oak dining room sets, 184–185
    oak timber production, 184, 186
    periodic income, 187
    sawtimber, 184–185
Tordon 101®, 141
Tordon 101R®, 141
Tordon RTU®, 141
TPA, *see* Trees per acre
Tractor-mounted sprayers, 145
Tree classes, 161–162, 175
Tree management
    butt log, hardwood tree, 173
    characteristics, 175–176
    cherrybark oak plantation, 171–172
    consulting forester/wildlife
        biologist, 180
    cooperative extension service, 179–180
    cull trees, 176
    cutting cycles, 168
    cutting stock, 176
    DBH, 171
    finance management operations, 168
    forestry management plan, 180–181
    hardwood species, 173–174
    hardwood tree quality, 170
    injury, hardwood, 173, 175
    internal decay damage, hardwood tree,
        173–174
    management potential, 170

mature hardwood forest, 167–168
natural regeneration harvest, 171–172
Northern red oak, productive upland
        site, 171
NRCS, 180
preferred and reserve
        growing stock, 176
stand age, 173
stand decision making, 168–169
state forestry agencies, 179
state wildlife agencies, 179
stocking guide, upland hardwood
        stands, 177
thinning, 178
timber production, 170
timber stand evaluation, 168–169
TPA, 176–177
vigor trees, 173–174
woods walk, 178–179
Tree species composition, 199–200
Trees per acre (TPA), 176–177

U

University forestry departments, 180
U.S. Department of Agriculture (USDA)
    oak/hickory forest, 25
    oak/pine forest types, 39
    online version, 87
    red maple, 29
    white oak/red oak/hickory forest, 28
    yellow poplar/white oak/northern red
        oak, 28

V

Vegetation and wildlife habitat
    canopy, 106–107
    composition and structure, 105
    food availability, 106–107
    soft mast production, 106
    sunlight, 107

W

Wetland Reserve Program (WRP), 247
White oak–black oak type group
    foot-slope positions, 201
    occurrence, 201

Ohio River region, 201
regeneration, 202
site, 200
White-tailed deer, 43–44
Wildlife habitat, 187–188
Wild turkeys, 116–117
Woodlands and savannahs
    blazing star, 226, 229
    bluestems, 226
    decimation, 224
    fire intensity and frequency, 226
    forbs, 226, 228
    grass and forb groundcover, 224
    grass-dominated groundcover, 224
    herbaceous ground layer, 225
    mesophication, 229
    midstories, 226, 230
    oak-shortleaf communities, 226
    Ohio Valley, 224
    park-like communities, 229
    prairie and forest, 224
    presettlement oak, 231
    red-cockaded woodpecker, 231
    ridges and rivers, 223
    site's productivity, 231
    transitional community, 241
    understory grasses, 226–227
WRP, *see* Wetland Reserve Program

Y

Young oak forests
    assessment and management
        diameter-limit, 221
        middle-aged category, 219
        natural disturbance, 220
        oak regeneration, 219
        oak stump sprouts, 221
        salvage operation, 220
    ecological succession, 218
    even-aged and uneven-aged, 254–255
    herbaceous groundcover, 219
    oak stump sprouts, 218
    rotation lengths, 255
    shade-intolerant shrubs, 219
    shelterwood, 254
    single-tree selection method, 255
    tops and branches, 219
    wildlife species, 218

Printed and bound by CPI Group (UK) Ltd, Croydon, CR0 4YY

17/10/2024

01775709-0005